T0219896

Practical Enterprise Data Lake Insights

Handle Data-Driven Challenges in an Enterprise Big Data Lake

Saurabh Gupta
Venkata Giri

Apress®

Practical Enterprise Data Lake Insights

Saurabh Gupta
Bangalore, Karnataka, India

Venkata Giri
Bangalore, Karnataka, India

ISBN-13 (pbk): 978-1-4842-3521-8
https://doi.org/10.1007/978-1-4842-3522-5

ISBN-13 (electronic): 978-1-4842-3522-5

Library of Congress Control Number: 2018948701

Managing Director, Apress Media LLC: Welmoed Spahr
Acquisitions Editor: Nikhil Karkal
Development Editor: Laura Berendson
Coordinating Editor: Divya Modi

Cover designed by eStudioCalamar

Cover image designed by Freepik (www.freepik.com)

Distributed to the book trade worldwide by Springer Science+Business Media New York, 233 Spring Street, 6th Floor, New York, NY 10013. Phone 1-800-SPRINGER, fax (201) 348-4505, e-mail orders-ny@springer-sbm.com, or visit www.springeronline.com. Apress Media, LLC is a California LLC and the sole member (owner) is Springer Science + Business Media Finance Inc (SSBM Finance Inc). SSBM Finance Inc is a **Delaware** corporation.

For information on translations, please e-mail rights@apress.com, or visit http://www.apress.com/rights-permissions.

Apress titles may be purchased in bulk for academic, corporate, or promotional use. eBook versions and licenses are also available for most titles. For more information, reference our Print and eBook Bulk Sales web page at http://www.apress.com/bulk-sales.

Any source code or other supplementary material referenced by the author in this book is available to readers on GitHub via the book's product page, located at www.apress.com/978-1-4842-3521-8. For more detailed information, please visit http://www.apress.com/source-code.

Printed on acid-free paper

Table of Contents

About the Authors

Saurabh Gupta is a technology leader, published author, and database enthusiast with more than 11 years of industry experience in data architecture, engineering, development, and administration. Working as a Manager, Data & Analytics at GE Transportation, his focus lies with data lake analytics programs that build digital solutions for business stakeholders. In the past, he has worked extensively with Oracle database design and development, PaaS and IaaS cloud service models, consolidation, and in-memory technologies. He has authored two books on advanced PL/SQL for Oracle versions 11g and 12c. He is a frequent speaker at numerous conferences organized by the user community and technical institutions. He tweets at @saurabhkg and blogs at sbhoracle.wordpress.com.

Venkata Giri currently works with GE Digital and has been involved with building resilient distributed services on a massive scale. He has worked on Bigdata tech stack, relational databases, high availability, and performance tuning. With over 20 years of experience in data technologies, he has in-depth knowledge of big data ecosystems, complex data ingestion pipelines, data engineering, data processing, and operations. Prior to GE, he worked with the data teams at LinkedIn and Yahoo.

About the Technical Reviewer

 As Director in LinkedIn's site reliability engineering organization, **Sai Selvaganesan** brings close to two decades of experience in data, from design, engineering, and operations to site reliability. With experience across multiple Silicon Valley companies including Apple, Yahoo, and LinkedIn, Sai's focus areas have been around scaling and optimizing data infrastructure and he holds multiple patents in the space.

Sai spearheaded strategic projects that helped forge multi-colo operations at LinkedIn. Previously, he worked on key initiatives including Yahoo's Panama project to overhaul search. Sai has a proven track record of building high-impact global teams focused on execution excellence and fuelling growth.

Sai holds a BA in Electrical Engineering from NIIT in India and is currently pursuing his MBA from UCLA.

Acknowledgments

We would like to thank Apress for giving us the opportunity to work on this project. A big shout goes out to the entire editorial team who have been extremely supportive throughout. Thanks Nikhil, Divya, and Laura. Trust me, it was not an episode, rather a journey.

Thanks Sai for accepting our request to review our content. It was indeed a great learning experience for us to have feedback from someone so humble and a master of the subject. We acknowledge your efforts in questioning us and ensuring quality of the product. We would like to graciously thank Janardh Bantupalli and Aditya for their distinguished contribution on change data capture and data operation topics.

Needless to say, all this would have never been possible without organizational support. Special thanks to GE legal for allowing us to pursue our interest. We would like to express our gratitude to Data & Analytics staff for their faith and encouragement. Thank you, Rick, Vijay, Libby, Jayadeep, Mayukh, and Diwakar.

Thanks to my family for bearing me all this time. It's not easy but whatever I am, is all because of your love and support. You are the life in me!

Foreword

When I was 10 years old, I would spend hours in the local library poring over books and recording pages and pages of notes, trying to soak up all the information I could. I was steadily building my knowledge bank so I would be ready with all the answers, whether I was applying that knowledge to write a book report or impress my parents with my rapid recall of statistics and facts about the world. I fast forward to today when my 8-year-old son calls out questions to the device on my kitchen counter and immediately gets answers, without having to access any websites, dig through books, or even leave his own house looking for that exact fact. In essence, learning from data that may be housed in a data lake instead of a structured data warehouse or in a book. The world has changed. We have volumes of data generated simply because of our ability to capture it – we are no longer limited to transactional systems or data captured only by written form. While the amount of data available is exponentially increasing, however, truly capitalizing on its value is dependent on having access when and how we need it. As technology leaders, we have the responsibility to make this data accessible so that it can be transformed into even more valuable information.

As a popularly covered topic in tech and management publications, some may ask, haven't we solved for that? Well, we've had a good start, but I would argue that new challenges have emerged. Information is not structured in the way it used to be instead it is being captured as both structured and unstructured data sets. As we lead our organizations forward, we must empower users through data democratization – putting the data in the hands of the end users so they can transform it into information in a relevant and meaningful way. The concept is powerful, and many organizations are embracing it, but the challenge of how to do it effectively remains a barrier. What are the stages of capturing the unstructured data, processing it and then allowing access to query it. On top of that, how do you manage the access and levels of security. These are challenging new questions that technology leaders face today.

The good news is that the challenges are not insurmountable. Importantly, though, is that, as the volume of data increases, the need to manage data processing with speed becomes paramount. Enterprise users have expectations of "consumer-like" experiences where speed and ease-of-use are key. What we need now is a practical approach to address this reality. From my experience, it starts with a cohesive enterprise data lake strategy. The data lake strategy needs to be architected with end user in mind and the opportunity to enable a variety of problem statements to be tackled. Unlike traditional transactional reporting where a problem statement is articulated at the beginning of the journey, the data lake attempts to fundamentally approach this in the inverse. Data is no longer a byproduct. Instead it is waiting for the user to apply a context and connect and discover data to convert it into information that can drive outcomes. The age of a data-driven culture has arrived and the principles and architecture of an enterprise data lake need to be ready to handle to volume, complexity, and flexibility.

Monica Caldas
CIO & SVP, GE Transportation
"Digital Leader of the Year" 2018
(http://womeninitawards.com/new-york/2018-usa-winners/)

CHAPTER 1

Introduction to Enterprise Data Lakes

"In God, we Trust; all others must bring data"

—*W. Edwards Deming*, a statistician who devised
"Plan-Do-Study-Act" method

It was in 1861 when Charles Joseph Minard, an 80-year-old French civil engineer, attempted to develop a visual that can narrate Napoleon's disastrous Russian campaign of 1812. Figure 1-1 depicts people movement and exhibits details on geography, time, temperature, troop count, course, and direction.

© Saurabh Gupta, Venkata Giri 2018
S. Gupta and V. Giri, *Practical Enterprise Data Lake Insights*,
https://doi.org/10.1007/978-1-4842-3522-5_1

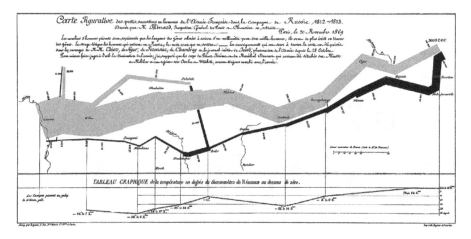

Figure 1-1. *Minard's map of Napolean's russing campaign in 1812*
Source: *"Worth a thousand words: A good graphic can tell a story,
bring a lump to the throat, even change policies. Here are three of
history's best." The Economist, December 19, 2007,* `https://www.`
`economist.com/node/10278643.`

From the above chart, in 1812, the Grand Army consisted of 422,000
personnel started from Poland; out of which only 100,000 reached Moscow
and 10,000 returned. The French community describes the tragedy as
"C'est la Bérézina".

The chart depicts the tragic tale with such clarity and precision. The
quality of the graph is accredited to the data analysis from Minard and a
variety of factors soaked in to produce high-quality map. It remains one of
the best examples of statistical visualization and data storytelling to date.
Many analysts have spent ample time to analyze through Minard's map
and prognosticated the steps he must have gone through before painting a
single image, though painful, of the entire tragedy.

Data analysis is not new in the information industry. What has grown
over the years is the data and the expectation and demand to churn
"gold" out of data. It would be an understatement to say that data has
brought nothing but a state of confusion in the industry. At times, data gets

unreasonable hype, though justified, by drawing an analogy with currency, oil, and everything precious on this planet.

Approximately two decades ago, data was a vaporous component of the information industry. All data used to exist raw, and was consumed raw, while its crude format remained unanalyzed. Back then, the dynamics of data extraction and storage were dignified areas that always posed challenges for enterprises. It all started with business-driven thoughts like variety, availability, scalability, and performance of data when companies started loving data. They were mindful of the fact that at some point, they need to come out of relational world and face the real challenge of data management. This was one of the biggest information revolutions that web 2.0 companies came across.

The information industry loves new trends provided they focus on business outcomes, catchy and exciting in learning terms, and largely uncovered. Big data picked up such a trend that organizations seemed to be in a rush to throw themselves under the bus, but failed miserably to formulate the strategy to handle data volume or variety that could potentially contribute in meaningful terms. The industry had a term for something that contained data: data warehouse, marts, reservoirs, or lakes. This created a lot of confusion but many prudent organizations were ready to take bets on data analytics.

Data explosion: the beginning

Data explosion was something that companies used to hear but never questioned their ability to handle it. Data was merely used to maintain a system of record of an event. However, multiple studies discussed the potential of data in decision making and business development. Quotes like "Data is the new currency" and "Data is the new oil of Digital Economy" struck headlines and urged many companies to classify data as a corporate asset.

Research provided tremendous value hidden in data that can give deep insight in decision making and business development. Almost every action within a "digital" ecosystem is data-related, that is, it either consumes or generates data in a structured or unstructured format. This data needs to be analyzed promptly to distill nuggets of information that can help enterprises grow.

So, what is Big Data? Is it bigger than expected? Well, the best way to define Big Data is to understand what traditional data is. When you are fully aware of data size, format, rate at which it is generated, and target value, datasets appear to be traditional and manageable with relational approaches. What if you are not familiar with what is coming? One doesn't know the data volume, structure, rate, and change factor. It could be structured or unstructured, in kilobytes or gigabytes, or even more. In addition, you are aware of the value that this data brings. This paradigm of data is capped as Big Data IT. Major areas that distinguish traditional datasets from big data ones are Volume, Velocity, and Variety. "Big" is rather a relative measure, so do the three "V" areas. Data volume may differ by industry and use case. In addition to the three V's, there are two more recent additions: Value and Veracity. Most of the time, the value that big data carries cannot be measured in units. Its true potential can be weighed only by the fact that it empowers business to make precise decisions and translates into positive business benefits. The best way to gauge ROI would be to compare big data investments against the business impact that it creates. Veracity refers to the accuracy of data. In the early stages of big data project lifecycle, quality, and accuracy of data matters to a certain extent but not entirely because the focus is on stability and scalability instead of quality. With the maturity of the ecosystem and solution stack, more and more analytical models consume big data and BI applications report insights, thereby instigating a fair idea about data quality. Based on this measure, data quality can be acted upon.

Let us have a quick look at the top Big Data trends in 2017 (Figure 1-2).

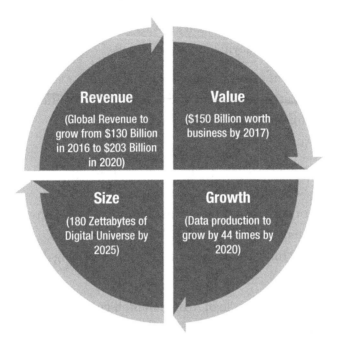

Figure 1-2. *Top big data trends in 2017. Source: Data from "Double-Digit Growth Forecast for the Worldwide Big Data and Business Analytics Market Through 2020 Led by Banking and Manufacturing Investments, According to IDC," International Data Corporation (IDC), October, 2016,* `https://www.idc.com/getdoc.jsp?container Id=prUS41826116`.

The top facts and predictions about Big Data in 2017 are:

1. Per IDC, worldwide revenues for big data and business analytics (BDA) will grow from $130.1 billion in 2016 to more than $203 billion in 2020.

2. Per IDC, the Digital Universe estimated is to grow to 180 Zettabytes by 2025 from pre-estimated 44 Zettabytes in 2020 and from less than 10 Zettabytes in 2015.

3. Traditional data is estimated to fold by 2.3 times between 2020 and 2025. In the same span of time, analyzable data will grow by 4.8 times and actionable data will grow by 9.6 times.

4. Data acumen continues to be a challenge. Organization alignment and a management mindset are found to be more business centric than data centric.

5. Technologies like Big Data, Internet of Things, data streaming, business intelligence, and cloud will converge to become a much more robust data management package. Cloud-based analytics to play key role in accelerating the adoption of big data analytics.

6. Deep Learning, one of Artificial Intelligence's (AI) strategies, will be a reality. It will be widely used for semantic indexing, and image and video tagging.

7. Non-relational analytical data stores will grow by 38.6% between 2015 to 2020.

Big data ecosystem

Big data IT strategy becomes critical when the nature of datasets goes beyond the capabilities of traditional (rather relational) approaches of handling data. At a high level, let us see what challenges Big Data brings to the table.

1. Data can be structured, semi-structured, or not structured at all. It is to impossible to design a generic strategy that can cater datasets of all structures.

2. Data from different sources can flow at different change rates. It may or may not have a schema.

3. How to process disparate datasets of sizes ranging from multi terabytes to multi petabytes together?

4. Common infrastructure must be cost effective and reliable, and should be fault tolerant and resilient. Total cost of ownership should be controllable to achieve high returns.

For a Big Data IT strategy to be successful, data must flow from a distinctive and reliable source system at a pre-determined frequency. Data must be relevant and mature enough to create critical insights and achieve specific business outcomes. From the cost perspective, enterprises were investing huge, in infrastructure to support storage, computing power, and parallelization.

Hadoop and MapReduce – Early days

In 2004, Google, in an effort to index the web, released white papers on data processing for large distributed data-intensive applications. The intent was to address two problem statements directly: storage and processing.

Google introduced MapReduce as the data processing framework and Google File System (GFS) as a scalable distributed file system. What makes the MapReduce framework highly scalable is the fact that a parallelized processing layer comes down all the way to the data layer that is distributed across multiple commodity machines. Google File System was designed for fault tolerance that can be accessed by multiple clients and achieve performance, scalability, availability, and reliability at the same time.

MapReduce proved to be the game changer for data process-intensive applications. It's a simple concept of breaking down a data processing task into a bunch of mappers that can run in parallel on thousands of commodity machines. Reducers constitute a second level of data processing operations that runon top of output generated from mappers.

Evolution of Hadoop

In the year 2002, the Yahoo! development team started a large-scale open-source web search project called Nutch. While Hadoop was still in the conceptual phase, Nutch's primary challenge was its inability to scale beyond a certain page limit. Then the concept of Google's GFS was introduced to project Nutch. A GFS-like file system resolved storage-related issues by allowing large files to sit in a system that was fault tolerant and available. By 2004, an open source implementation of GFS was ready as Nutch Distributed Filesystem (NDFS).

In 2004, Google introduced the MapReduce processing framework, which for obvious reasons, was immediately added into project Nutch. By early 2005, Nutch algorithms were already working with NDFS and MapReduce at an enterprise level. Such instrumentation was the combination of NDFS and MapReduce that, in 2006, Yahoo! took this package out of project Nutch. Doug Cutting was fascinated by a little stuffy yellow elephant and named this package Hadoop for the ease of memory and pronunciation. In 2008, Apache Software Foundation took over Hadoop to work beyond web-search optimization and indexing.

In a series of events starting in 2008, Hadoop stack has been pulling some magical numbers to prove its power of processing and worth at the enterprise level. In February 2008, Yahoo! claimed to generate a web search index on 10,000 core Hadoop cluster. In April 2008, Apache Hadoop set a world record as the fastest platform to process terabyte of data with a 910-node cluster. Hadoop could sort one terabyte of data in just 209 seconds, beating the previous benchmark of 297 seconds.

Hadoop 1.0 was introduced by end of the year 2011. The basic flavor of Hadoop focused on providing the storage and processing framework. The concept, MapReduce processing coupled with Hadoop Distributed Filesystem (HDFS), gained wide traction and quick adoption in the industry. Though this setup was largely appreciated due to flexibility and ease of implementation, concerns over resource management, scalability, security, and availability were still on. These drawbacks restricted the enterprise level adoption of HDFS. A high-level architecture of Hadoop 1 exhibits key components of HDFS storage layer and MapReduce processing layer. (Figure 1-3).

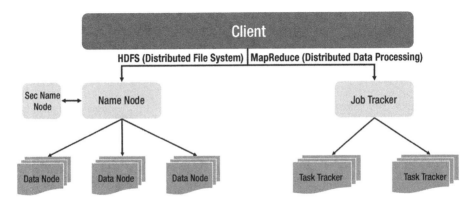

Figure 1-3. *Hadoop 1 high-level architecture*

Later in 2013, Hadoop 2.0 came out with brand new features that addressed availability and security. However, the major component in Hadoop 2.X was YARN (Yet Another Resource Negotiator). Resource management in Hadoop 1.x used to be carried out by a job tracker. Hadoop 2.x lays down another layer for resource management through YARN and segregates load management from job execution. YARN becomes responsible for resource allocation for all operations within the cluster. The MapReduce operation runs in a shell called Application Master who seeks and receives resources through YARN. It is backward compatible with Hadoop 1.x as well. Figure 1-4 positions storage and processing components of Hadoop 2. Key callouts from the below architecture are:

- Standby NameNode to support high availability of primary NameNode

- YARN for cohesive resource management and efficient job scheduling

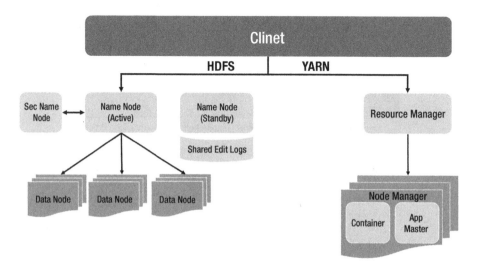

Figure 1-4. *Hadoop 2 high-level architecture*

Figure 1-5 highlights the difference between Hadoop 1.x and Hadoop 2.x at the skeleton level.

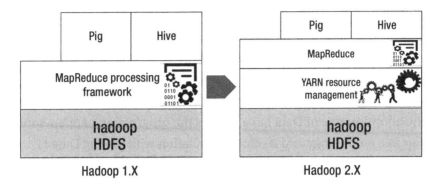

Figure 1-5. *Head to head comparison of Hadoop 1 and Hadoop 2*

History of Data Lake

Since the time Big Data trends have become buzzwords, several marketing terms have been coined to describe data management strategies. Eventually, all of them happen to represent a version of the Big Data ecosystem.

In 2010,[1] James Dixon came up with a "time machine" vision of data. Data Lake represents a state of enterprise at any given time. The idea is to store all the data in a detailed fashion in one place and empower business analytics applications, and predictive and deep learning models with one "time machine" data store. This leads to a Data-as-an-Asset strategy wherein continuous flow and integration of data enriches Data Lake to be thick, veracious, and reliable. By virtue of its design and architecture, Data Lake plays a key role in unifying data discovery, data science, and enterprise BI in an organization.

---

[1]Woods, Dan; "James Dixon Imagines a Data Lake that Matters," Forbes, `https://www.forbes.com/sites/danwoods/2015/01/26/james-dixon-imagines-a-data-lake-that-matters/#1dd2c5e34fdb`

According to James Dixon[2] *"If you think of a datamart as a store of bottled water—cleansed and packaged and structured for easy consumption—the data lake is a large body of water in a more natural state. The contents of the data lake stream in from a source to fill the lake, and various users of the lake can come to examine, dive in, or take samples."*

The adoption rate of Data Lake reflects the progression of open-source Hadoop as a technology and its close association with the Big Data IT trend. Keep in mind that although data lake is presumed to be on Hadoop due to the latter's smooth equation with Big Data IT, it's not mandatory. Don't be surprised if you Find Data Lake hosted on relational databases. You must factor in the cost of standing a non-Hadoop stack, data size, and BI use cases.

Data Lake: the concept

Data Lake is a single window snapshot of all enterprise data in its raw format, be it structured, semi-structured, or unstructured. Starting from curating the data ingestion pipeline to the transformation layer for analytical consumption, every aspect of data gets addressed in a data lake ecosystem. It is supposed to hold enormous volumes of data of varied structures.

Data Lake is largely a product that is built using Hadoop and a processing framework. The choice of Hadoop is as direct as it can be. It not only provides scalability and resiliency, but also lowers the total cost of ownership.

At a broader level, data lake can be split into a data landing layer and an analytical layer. From the source systems, data lands directly into the data landing or mirror layer. The landing layer contains the as-is copy of the data from source systems, that is, raw data. It lays the foundation for the battleground for the analytical layer. The analytical layer is a

---

[2]https://jamesdixon.wordpress.com/2010/10/14/pentaho-Hadoop-and-data-lakes/

highly dynamic one in the Data Lake world as it is the downstream consumer of raw data from the mirror layer. The landing or mirror layer data is fed through a transformation layer and builds up the analytical or consumption layer. The analytical layer ensures data readiness for data analytics sandbox and thus, acts as a face-off to data scientists and analysts. Pre-built analytical models can be directly plugged in to run over the consumption layer. Perhaps, dynamic analytics like data discovery or profiling models can also be made to run directly on the consumption layer. Data visualization stacks can consume data from the consumption layer to present key indicators and data trends.

Another split of data lake can be based on temporal dimensions of data. Historical raw data can be archived and stored securely within the data lake. While it will still be active to the data lake consumers, it can be moved to a secondary storage. Mirror layer that we discussed above can hold incremental data given a pre-determined timeline. In this case, consumption layer is built upon augmented mirror layer only.

This model doesn't need to have physical data marts that are custom built to serve a singular static model. Rather they transform the data in a usable format to enable analytics and business insights. On the other hand, data warehouse provides an abstract image of a specific business wing.

Data lake architecture

Cost and IT simplification are the biggest features of Data Lake. Inexpensive Hadoop storage with schema-less-write capability and in-house processing framework using hive, pig, or python largely the success of data lake. Figure 1-6 lays out a high-level wireframe of an enterprise Data Lake.

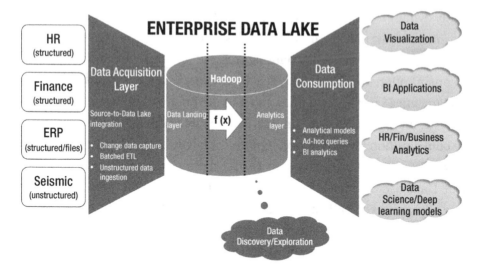

Figure 1-6. *Enterprise Data Lake Architecture*

In the above architecture diagram, there may be multiple source systems dumping data into the Enterprise Data Lake. Source systems can be of variety of nature and structure. It may come from relational sources, static file systems, web logs, or time-series sensor data from Internet of Things devices. It may or may not be structured. Without hampering the structure of data from the source and without investing into data modeling efforts in Hadoop, all source systems ingest data into Data Lake in stipulated real time.

Once data comes in the purview of Data Lake frontiers, it propagates through the processing layer to build the analytical layer. At this stage, it may be required to define the schema and structure for raw data. Thereafter, depending upon the data exchange guidelines laid down by the data governance council, data gets consumed by predictive learning models, BI applications, and data science tracks. Meanwhile, the anatomy of data discovery continues to provide a visual and exploratory face to big data in Hadoop Data Lake by directly working on raw data.

Why Data Lake?

Big Data IT is driven by competition. Organizations want to exploit the power of data analytics at a manageable cost to stand out to their competitors. Data Lake provides a scalable framework to store massive volumes of data and churn out analytical insights that can help them in effective decision making and growing new markets. It brings in a shift of focus from protracted and laborious infrastructure planning exercise to data-driven architecture. Data ingestion and processing framework becomes the cornerstone rather than just the storage.

Another perspective that comes with Data Lake building is the simplified infrastructure. Organizations spend a lot in building reliable stack for different nature of data and usually follow best fit approach to manage data. For example, relational databases have been the industry de-facto for structured data for ages. For the semi-structured and unstructured data coming through sensors, web logs, social media, traditional file systems were being used widely. At the end of the day, they have data marts bracketed by data structure, use cases, and customer needs but incur high capital and operational expenses. All such scenarios are easily addressed with Data Lake with a simplified infrastructure.

Data Lake induces accessibility and catalyzes availability. It warrants data discovery platforms to soak the data trends at a horizontal scale and produce visual insights. It largely cuts down the time that goes into data preparation and exhaustive data analysis. Figure 1-7 shows the key attributes of Data Lake.

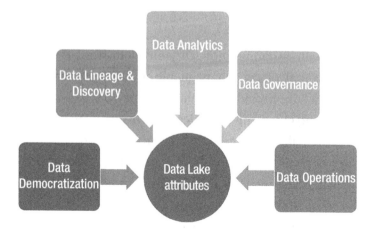

Figure 1-7. *Data Lake attributes*

Data Lake Characteristics

Let us understand key characteristics of an enterprise data lake.

1. Data lake must be built using a scalable and fault
    tolerant framework – the data lake concept focusses
    upon simplification and cost reduction without
    compromising on quality and availability. Apache
    Hadoop provides cost benefits by running on
    commodity machines and brings resiliency and
    scalability as well.

2. Availability – data in the data lake must be accurate
    and available to all consumers as soon as it is
    ingested.

3. Accessibility – shared access models to ensure
    data can be accessed by all applications. Unless
    required at the consumption layer, data shards are
    not a recommended design within the data lake.

Another key aspect is data privacy and exchange regulations. Data governance council is expected to formulate norms on data access, privacy and movement.

4. Strategy to track data lineage, right from the source systems up to consumption – the data lineage tracker provides a single snapshot of life cycle of data. Starting from its source onset, the tracker would depict data's movement and consumption through layers and applications.

5. Data reconciliation strategy from the source systems – from the data operations perspective, data reconciliation is a critical facet of quality.

6. Data governance policies must not enforce constraints on data – Data governance intends to control the level of democracy within the data lake. Its sole purpose of existence is to maintain the quality level through audits, compliance, and timely checks. Data flow, either by its size or quality, must not be constrained through governance norms.

7. Data in the data lake should never get disposed. Data driven strategy must define steps to version the data and handle deletes and updates from the source systems.

8. Support for in-place data analytics – the data lake is a singular view from all the source systems to empower in-house data analytics. Downstream applications can extract the data from the consumption layer to feed a disparate application.

9. Data security – security is a critical piece of
    Data Lake. Enterprises can start with a reactive
    approach and strive for proactive ways to detect
    vulnerabilities. Data-centric security models should
    be capable of building real-time risk profiles that can
    help detect anomalies in user access or privileges.

10. Archival strategy – as part of ILM strategy
    (Information Lifecycle Management), data retention
    policies must be created. Retention factor of data,
    that resides in a relatively "cold" region of lake, must
    be considered, which is not a big deal in the Hadoop
    world though, but storage consumption multiplied
    by new data exploration, brings a lot of wisdom to
    formulate a data archival strategy.

Figure 1-8 draws out the steps involved in the data lake planning and construction exercises. In the construction phase, you would note that tasks are more data centric compared to the ones in the planning phase.

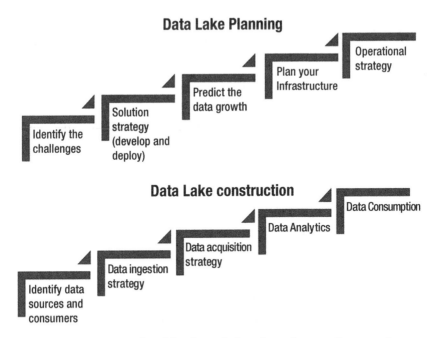

Figure 1-8. *Steps involved in data lake planning and execution*

Data lake vs. Data warehouse

Data warehousing, as we are aware, is the traditional approach of consolidating data from multiple source systems and combining into one store that would serve as the source for analytical and business intelligence reporting. The concept of data warehousing resolved the problems of data heterogeneity and low-level integration. In terms of objectives, a data lake is no different from a data warehouse. Both are primary advocates of terms like "single source of truth" and "central data repository."

In basic terms, a data lake ingests all data in its raw format, unprocessed and untouched to build a huge data store. It goes super trendy with Hadoop due to its sheer volume and affinity for distributed computing. A data warehouse, on the other hand, extracts data from the source systems that pass through a processing layer before settling down in different schemas (schema on-write). All data in a warehouse is processed, well modeled, and structured. In fact, the building paradigm that both the stores follow can be closely associated to its target users. Data warehouse targets business professionals, management, and business analysts who expect structured analysis reports at the end of the day. On the contrary, a data lake opens a war room for data scientists, analysts, and data engineering specialists from multiple domains for data crunching, exploration, and refining.

Open-source Hadoop cannot be the first choice to build a data warehouse. A data warehouse runs on relatively expensive storage to withstand high-scale data processing. Data lake is primarily hosted on Hadoop, which is open source and comes with free community support. Total cost of ownership and return on investments need to be factored in while comparing a data warehouse and a data lake. Figure 1-9 shows use case differences between warehouses and data lakes.

Figure 1-9. *Data warehouse versus Data Lake*

A data warehouse follows a pre-built static structure to model source data. Any changes at the structural and configuration level must go through a stringent business review process and impact analysis. Data lakes are very agile. Consumption or analytical layer can be modified to fit in the model requirements. Consumers of a data lake are not constant; therefore, schema and modeling lies at the liberty of analysts and scientists.

Various analogies have been drawn to clarify the difference between the two worlds. What differentiates a Cricket World Cup from the Olympics is exactly how a warehouse is different from a data lake.

How to achieve success with Data Lake?

The strategy to hold the data and extract its real value plays a vital role in justifying the capital and operational expenses incurred upon building up the data lake. There might be questions around data classification and its criticality quotient. How long is the data valuable to a business? What's the measure of potential in a data slice? Let us check out a few key drivers that can ensure a data lake's success.

Facebook works with 30+ Petabytes of user-generated data. Google is concerned with every bit that we do in a day. Data-driven companies never shy away from gathering and storing data. Vision is loud and clear – data has no expiration date. Every bit of data carries a value that can be maximized. Therefore, you must have a clear understanding of who the consumers are, what use cases are served by the data lake currently, and most importantly, what's the vision?

Vision drives a lot of data strategies. Data late curators need to understand what is being consumed and what is needed. First, the pace at which data acquisition occurs determines the measure of data lake enrichment as an asset. This lifts the data ingestion barriers, allowing data to grow centrally and restrict duplication and proliferation of data. Second, the rate at which raw data gets consumed indicates the measure at which it becomes dormant. This will push data stewards to rethink data archival strategy and create space for fresh data from existing and new sources.

Data governance and data operations

Data-driven enterprises are emerging as ardent agile practitioners and DevOps transformers for one common objective. The code deployment process becomes faster and enables business users to analyze the impact of "tiny" incremental fixes. Facebook adopted a "quasi-continuous release cycle" to eliminate the need for hotfixes, empower global engineering team to support development and deployment, and quantify the user experience.

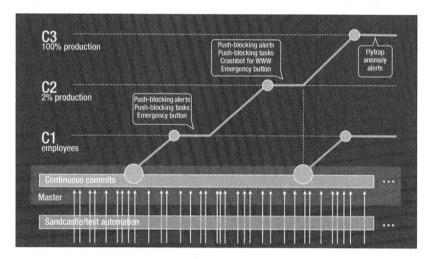

Figure 1-10. *Rapid release at massive scale. Source: Rossi, Chuck; "Rapid Release at Massive Scale," August 31, 2017,* `https://code.` `facebook.com/posts/270314900139291/rapid-release-at-` `massive-scale/`

Data Lake, in its initial stages, acts as a playgroup to try out pre-built analytical models that are well versed with the resident data. As data lake matures to an instrumental level, it opens a war room for data scientists and analysts to be ruthless with data churning exercise.

Situations like this give a glimpse of the pace at which "changes" move. And this calls out for a need of an administrative body that can operationalize these changes with negligible impact. The framework, known as data operations, becomes extremely critical to ensure smooth data lake functions in terms of availability, performance, and security. Data operations eventually become the gatekeeper of the data lake with respect to code deployments, platform upgrades, and application support. All code changes officiating for production promotion pass through the operations process layer. Key deliverables of data operations are listed below.

1. DevOps handshake – Data lake operations desk accompanies DevOps team at critical stages of development and deployment

2. Availability and maintenance – Operations shoulder the responsibility of ensuring availability of the data lake infrastructure. They plan and execute platform upgrades and coordinate periodic maintenance activities.

3. Release management – Codebase maintenance and production releases are guarded by data lake operations.

4. Monitoring of data flows and capture key data lake metrics – Data lake operations employ a layered support model for continuous monitoring of data movement from mirror to analytical and further consumption.

5. Runbook and standard operating procedures – Documents that describe formal and ad-hoc exercises to be carried out by data lake operations.

When Facebook acquired WhatsApp, both had different privacy norms, which further called for an investigation on data sharing. The Electronic Privacy Information Center and the Center for Digital Democracy filed a complaint with the Federal Trade Commission (FTC) stating "Specifically,

WhatsApp users could not reasonably have anticipated that by selecting a pro-privacy messaging service, they would subject their data to Facebook's data collection practices," leading to violation of Section 5 of the FTC Act, 15 U.S.C. § 45(n). It led to the revision of data governance laws at Facebook.

Figure 1-11 shows the tasks carried out by data governance and data operations at different layers of an enterprise data lake.

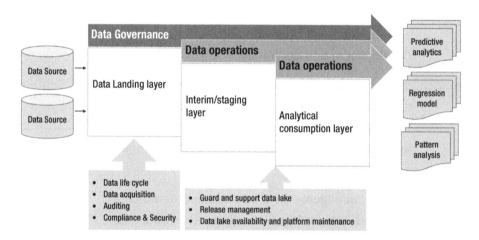

Figure 1-11. *Concentration of data governance and data operations in a data lake*

The data governance council comprises key stakeholders of the data who are aware with the nuances of data privacy, engineering, processing, and movement. In its document form, it can be a questionnaire before the team promotes a change or a fix to the data lake. Questionnaires or rules may include a run-book for the change, risk evaluation metric, impact analysis, and sign-off from the key stakeholders. The council can also audit the data lake to keep track of changes being made in a period rather than just at deployment. The below list builds the portfolio of an effective data governance policy.

1. Data Acquisition – Formulate a strategy to idealize the data lake as an asset. This piece of governance council defines what organizational data can be lodged within the lake. When should it come and how?

2. Data catalog – Cataloging is a critical piece when explicating data lake roadmap and vision. Data scientists and analyst can be crucial in providing feedback and proposing council on what data could be vital for effective analytics.

3. Data organization – The governance council should also focus on structure and format of data to enrich and maintain synergy within the data lake.

4. Metadata management – Data governance stewards keep an eye on data quality, profiling, and lineage. It helps in the evolution of data categories like internal, confidential, public, and others.

5. Compliance and security – Deal with data security, access management, and abide by the organization's compliance policies.

Effective data governance elevates confidence in data lake quality and stability, which is a critical factor to data lake success story. Data compliance, data sharing, risk and privacy evaluation, access management, and data security are all factors that impact regulation.

Data democratization with data lake

One of the appreciative traits of a data warehouse is that each ETL process or development holds a definite business objective. Every penny of data in a warehouse undergoes a thorough validation and approval process.

On the other hand, data lake has been an ardent supporter of data democratization.

To withstand today's data-driven economy, organizations tend to work with lots of data. The datasets exist in different shapes and sizes and may follow different routes of consumption. However, this data is discoverable to only those who are familiar with data lineage. Non-data practitioners may find it tedious to explore and play with data of interest. Half of their time is wasted in data discovery, data cleansing, and hunting down reliable sources of data. Data dwelling in silos prevents liberation of data to its full potential.

Data democratization is the concept of diluting data isolation and ensuring that data is seamlessly available to the appropriate takers on time. In addition to data agility and reliability, data democratization lays down a layer of data accessibility that helps in discovering data quickly through custom tools and technologies. The objective is to empower data analysts with swift access to the veracious data set, so as to enable rapid data analysis and decision making.

Let us discuss a few of the approaches.

1. Curated data layers – Curated data layers help in flattening out data models for functional users. These users may be interested in only 10% of total raw information but could struggle in narrowing down the data of interest. The objects contained in curated layers are intended to provide sliced and diced data in a flattened-out structure. Figure 1-12 shows the SALES curated layer built on top of six base layer tables.

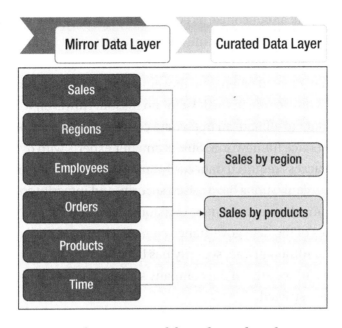

Figure 1-12. *SALES layer curated from base data layer*

2. Self-service platforms – Self-service portals may
    act as a data marketplace wherein a user can
    traverse through data sets, discover based on
    functional implications, mold and transform data
    representations, and download them for personal
    records. In the backend, the framework can work
    through with APIs for dataset transformation, data
    discovery, and extraction.

By collating all the key elements, data democratization cuts across the
concepts of *data nirvana* and *open data* paradigms. Data democratization
can be achieved in an enterprise data lake. The basic idea behind data lake
is to treat all the data equally relevant and insource into a single platform
without the barriers of architecture, models, or predefined framework.

Let us now talk about concerns connected to data democratization. An argument that has gained community support states that democratization catalyzes the ability of a data lake to become a data swamp. Keep in mind that swamps are created when data movement is unregulated and a lot of irrelevant data houses are in the lake for no reason. However, we need to democratize data to affiliate an open data concept by easing data discovery through self-service frameworks. Subject matter experts with diversified expertise can access required data easily and contribute to organizational insights. Few organizations have raised security and integrity concerns of democratizing data for the entire workforce. This adds to the responsibilities of the data governance council who are in charge of data categorization. Although data security has been on the rise over the last few years, the surface area of the company's internal and confidential data should still be restricted.

Fast Data - Life beyond Big Data

Today, Big Data analytics is popular. Enterprises not into data analytics are extinct or endangered. While there are many still getting their feet wet, there were organizations who not only achieved "big data" milestone well before, but also came up with cases that demanded stretched-out capabilities of big data.

One of the facts that we realized in this chapter is that Big Data is not just "Big", but it is fast, precious, and relatively mysterious. Big Data is an abstract trend while Data Lake is an ecosystem that implements big data analytics. Data from the source systems, in its original format, be it structured or unstructured, flows into the data lake at different magnitudes. It may be in the scale of gigabytes per second or terabytes per hour. However, the data settles down first and then gets processed to build an analytical layer. Consumption models are oblivious to the fact that what they consume has gone through levels of functional and logical transformation. A lineage tracking exercise will reflect the lag between the data generation phase and the consumption phase. This lag is accredited to the ETL pipeline in the data acquisition layer and the time consumed during transformation, which leads to delayed analytics. Let us see a few use cases.

A telecom company monitors all international calls in the country. It applies the NLP algorithm on voice intercepts to filter out suspicious calls and send out a notification to the security agencies. The expectation is to have monitoring in real time and any delayed alert will be treated as an opportunity missed. Although data volume and velocity are high, ingestion and analysis are critical.

Another scenario could be a cricket analytics website that needs to predict if a batsman will be bowled or run out on the next ball (depending on his current as well as last 10 outings). Cases like these share a common trait. It is not the data size that matters but what matters is how fresh the data is at the time of producing actionable insights.

Use cases like those described above have led to the emergence of a revolution known as Fast Data Analytics. Fast data takes a leap ahead of Big Data and strives for "fresh" data for mining. While Big Data struggled with the challenges of distributed storage and computing, fast data focuses on the time-sensitivity of data for analytics. Hadoop provided a batch processing platform but it's inability to process data in near real-time or real-time has been under scrutiny. Nevertheless, real-time processing will be difficult for Hadoop. Fast data analytics works on the data as soon as it gets into the data lake.

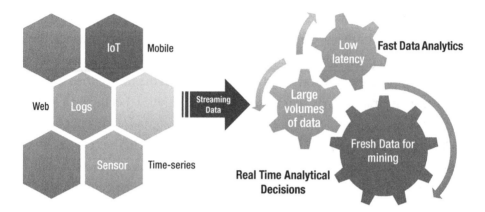

Figure 1-13. *Fast data analytics is the next gen analytics engine where it focuses on time-sensitivity factor of data for analytics*

What fast data analytics achieves is proactive real-time insights. It can play a significant role in reducing latency, enhancing accuracy, faster decision making, and improved customer experience. Applications that leverage data from sensors and machine to machine communication in the Internet of Things provide the best fit use case for fast data analytics.

Conclusion

Let us quickly revisit the concepts covered to introduce a data lake. Big data can be structured or unstructured and is relatively big in terms of volume, velocity, and value. Market analysts predicted big data to be a key aspect to several business insights, provided it is appropriately stored and analyzed. The Organizations realize the value hidden in this heap but huge investments in storage and building efficient computing models was a challenge. In 2004, Google's concept of distributed file systems and distributed processing frameworks comes to the rescue. The evolution of the Hadoop distributed file system has been a rapid expedition as it enabled storage on relatively cheap commodity hardware. MapReduce, the distributed computing framework, instantiated processing framework to connect with data for processing.

Data lake is an ecosystem for the realization of big data analytics. What makes data lake a huge success is its ability to contain raw data in its native format on a commodity machine and enable a variety of data analytics models to consume data through a unified analytical layer. While the data lake remains highly agile and data-centric, the data governance council governs the data privacy norms, data exchange policies, and the ensures quality and reliability of data lake.

While big data trends on a run, fast data analytics is the one that is picking up the speed lately. It is different from big data and deals with the same volumes and structures of data but endeavors data freshness more than data volume. Recently, enterprises have started realizing the true essence of data usage. All big data cannot be fast and at the same time, all fast data need not be big. The data lake ecosystem can complement the two concepts entirely and empower organizations with both big and fast flavors of data.

In the next chapter, we will start off with a deep dive into data ingestion principles. We will understand batched ingestion, real-time data movement, and change data capture concepts along with architecture and design considerations.

CHAPTER 2

Data lake ingestion strategies

"If we have data, let's look at data. If all we have are opinions, let's go with mine."

—*Jim Barksdale, former CEO of Netscape*

Big data strategy, as we learned, is a cost effective and analytics driven package of flexible, pluggable, and customized technology stacks. Organizations who embarked into Big Data world, realized that it's not just a trend to follow but a journey to live. Big data offers an open ground of unprecedented challenges that demand logical and analytical exploitation of data-driven technologies. Early embracers who picked up their journeys with trivial solutions of data extraction and ingestion, accept the fact that conventional techniques were rather pro-relational and are not easy in the big data world. Traditional approaches of data storage, processing, and ingestion fall well short of their bandwidth to handle variety, disparity, and volume of data.

In the previous chapter, we had an introduction to a data lake architecture. It has three major layers namely data acquisition, data processing, and data consumption. The one that is responsible for building and growing the data lake is the data acquisition layer. Data acquisition lays the framework for data extraction from source data systems and

© Saurabh Gupta, Venkata Giri 2018
S. Gupta and V. Giri, *Practical Enterprise Data Lake Insights*,
https://doi.org/10.1007/978-1-4842-3522-5_2

orchestration of ingestion strategies into data lake. The ingestion framework plays a pivotal role in data lake ecosystem by devising data as an asset strategy and churning out enterprise value.

The focus of this chapter will revolve around data ingestion approaches in the real world. We start with ingestion principles and discuss design considerations in detail. The concentration of the chapter will be high on fundamentals and not on tutoring commercial products.

What is data ingestion?

Data ingestion framework captures data from multiple data sources and ingests it into big data lake. The framework securely connects to different sources, captures the changes, and replicates them in the data lake. The data ingestion framework keeps the data lake consistent with the data changes at the source systems; thus, making it a single station of enterprise data.

A standard ingestion framework consists of two components, namely, *Data Collector* and *Data Integrator*. While the data collector is responsible for collecting or *pulling* the data from a data source, the data integrator component takes care of ingesting the data into the data lake. Implementation and design of the data collector and integrator components can be flexible as per the big data technology stack.

Before we turn our discussion to ingestion challenges and principles, let us explore the operating modes of data ingestion. It can operate either in real-time or batch mode. By virtue of their names, real-time mode means that changes are applied to the data lake as soon as they happen, while a batched mode ingestion applies the changes in batches. However, it is important to note that real-time has its own share of lag between change event and application. For this reason, real-time can be fairly understood as near real-time. The factors that determine the ingestion operating mode are data change rate at source and volume of this change. Data change rate is a measure of changes occurring every hour.

For real-time ingestion mode, a change data capture (CDC) system is sufficient for the ingestion requirements. The change data capture framework reads the changes from transaction logs that are replicated in the data lake. Data latency between capture and integration phases is very minimal. Top software vendors like Oracle, HVR, Talend, Informatica, Pentaho, and IBM provide data integration tools that operate in real time.

In a batched ingestion mode, changes are captured and persisted every defined interval of time, and then applied to data lake in chunks. Data latency is the time gap between the capture and integration jobs. Figure 2-1 illustrates the challenges of building an ingestion framework.

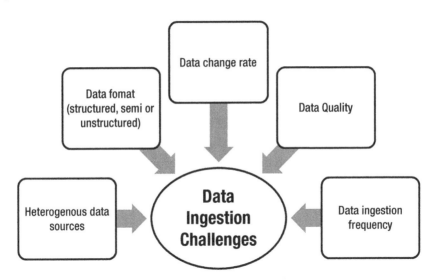

Figure 2-1. *Data Ingestion challenges*

Understand the data sources

Selection of data sources for data lake is imperative while enriching analytical acumen for a business statement. Data sources form the basis of the data acquisition layer of a data lake. Let us look at the variety of data sources that can potentially ingest data into a data lake.

- OLTP systems and relational data stores – structured data from typical relational data stores can be ingested directly into a data lake.

- Data management systems – documents and text files associated with a business entity. Most of the time, these are semi-structured and can be parsed to fit in a structured format.

- Legacy systems – essential for historical and regulatory analytics. Mainframe based applications, customer relationship management (CRM) systems, and legacy ERPs can help in pattern analysis and building consumer profiles.

- Sensors and IoT devices – devices installed on healthcare, home, and mobile appliances and large machines can upload logs to a data lake at periodic intervals or in a secure network region. Intelligent and real-time analytics can help in proactive recommendations, building health patterns, and surmising meteoric activities and climatic forecast.

- Web content – social media platforms like Facebook, Twitter, LinkedIn, Instagram, and blogs accumulate humongous amounts of data. It may contain free text, images, or videos that is used to study user's behavior, business focused profiles, content, and campaigns.

- Geographical details – data flowing from location data, maps, and geo-positioning systems.

Structured vs. Semi-structured vs. Unstructured data

Data serves as the primitive unit of information. At a high level, data flows from distinct source systems to a data lake, goes through a processing layer, and augments an analytical insight. This might sound quite smooth but what needs to be factored in is the data format. Data classification is a critical component of the ingestion framework. Data can be either structured, semi-structured, or unstructured. Depending on the structure of data, the processing framework can be designed effectively.

Structured data is an organized piece of information that aligns strongly with the relational standards. It can be searched using a structured query language and the result containing the data set can be retrieved. For example, relational databases predominantly hold structured data. The fact that structured data constitutes a very small chunk of global data cannot be denied. There is lot of information that cannot be captured in a structured format.

Unstructured data is the unmalleable format of data. It lacks a structure; thus, making basic data operations like fetch, search, and result consolidation quite tedious. Data sourced from complex source systems like web logs, multimedia files, images, emails, and documents are unstructured. In a data lake ecosystem, unstructured data forms a pool that must be wisely exploited to achieve analytic competency. Challenges come with the structure and volume. Documents in character format (text, csv, word, XML) are considered as semi-structured as they follow a discernable pattern and possess the ability to be parsed and stored in the database. Images, emails, weblogs, data feeds, sensors, and machine-generated data from IoT devices, audio, or video files exist in binary format and it is not possible for structured semantics to parse this information.

"Unstructured information represents the largest, most current, and fastest growing source of knowledge available to businesses and governments. It includes documents found on the web, plus an estimated 80% of the information generated by enterprises around the world." - Organization for the Advancement of Structured Information Standard (OASIS) - a global nonprofit consortium that works towards building up the standards for various technology tracks (`https://www.oasis-open.org/`).

Each of us generate a high volume of unstructured data every day. We are connected to the web every single hour as share data in one or the other way via a handful of devices. The amount of data we produce on social media or web portals gets proliferated to multiple downstream systems. Without caring much, we shop for our needs, share what we think, and upload files to share. By data retention norms, data never gets deleted but follows the standard information lifecycle management policy set by the organization. At the same time, let's be aware that information baked inside unstructured data files can be enormously useful for data analysis. Figure 2-2 lists the complexities of handling unstructured data in the real world. Data without structure and metadata is difficult to comprehend and fit into pre-built models.

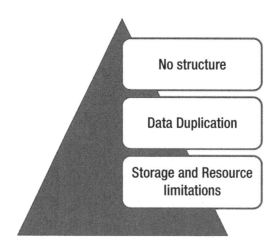

Figure 2-2. *Unstructured data complexities*

Apache Hadoop is a proven platform that addresses the challenges of unstructured data in the following ways:

1. Distributed storage and distributed computing – Hadoop's distributed framework favors storage and processing of enormous volumes of unstructured data.

2. Schema on read – Hadoop doesn't require a schema on write for unstructured data. It is only post processing that analyzed data needs a schema on read.

3. Complex processing – Hadoop empowers the developer community to program complex algorithms for unstructured data analysis and leverages the power of distributed computing.

Data ingestion framework parameters

Architecting data ingestion strategy requires in-depth understanding of source systems and service level agreements of ingestion framework. From the ingestion framework SLAs standpoint, below are the critical factors.

- Batch, real-time, or orchestrated – Depending on the transfer data size, ingestion mode can be batch or real time. Under batch mode, data movement will trigger only after a batch of definite size is ready. If the data change rate is defined and controllable (such that latency is not impacted), real-time mode can be chosen. For incremental change to apply, ingestion jobs can be orchestrated at periodic intervals.

- Deployment model (cloud or on-premise) – data lake can be hosted on-premise as well as public cloud infrastructures. In recent times, due to the growing cost of computing and storage systems, enterprises have started evaluating cloud setup options. With a cloud hosted data lake, total cost of ownership (TCO) decreases substantially while return on investment (ROI) increases.

An ingestion strategy attains stability only if it is able enough to handle disparate data sources. The following aspects need to be factored in while understanding the source systems.

- Data lineage – it is a worthwhile exercise to maintain a catalog of the source systems and understand its lineage starting from data generation until the ingestion entry point. This piece could be fully owned by the data governance council and may get reviewed from time to time to align and cover catalog registrants under the ongoing compliance regulations.

- Data format – whether incoming data is in the form of data blocks or objects (semi or unstructured)

- Performance and data change rate – data change rate is defined as the size of changes occur every hour. It helps in selecting the appropriate ingestion tool in the framework.

- Performance is a derivative of throughput and latency.

- Data location and security

  - Whether data is located on-premise or in a public cloud infrastructure, network bandwidth plays an important role.

  - If the data source is enclosed within a security layer, the ingestion framework should be enabled and establishment of a secure tunnel to collect data for ingestion should occur.

- Transfer data size (file compression and file splitting) – what would be the average and maximum size of block or object in a single ingestion operation?

- Target file format – Data from a source system needs to be ingested in a Hadoop compatible file format.

Table 2-1 compiles the list of file formats, their features, and scenarios in which they are preferred for use.

Table 2-1. *File formats and their features*

File type	Features	Usage
Parquet	• Columnar data representation • Nested data structures	• Good query performance • Hive supports schema evolution • Optimized for Cloudera Impala • Slower write performance
Avro	• Row format data representation • Nested data structures	• Stores metadata • Supports file splitting and block compression
ORC	• Optimized Record Columnar files • Row format data representation as key-value pair • Hybrid of row and columnar format • Row format helps to keep data intact on the same node • Columnar format yields better compression	• Good for data query operations • Improved compression • Slow write performance • Schema evolution not supported • Not supported by Cloudera Impala
SequenceFile	• Flat files as key-value pairs	• Limited schema evolution • Supports block compression • Used as interim files during MapReduce jobs
CSV or Text file	• Regular semi-structured files	• Easy to be parsed • No support for block compression • Schema evolution not easy

(continued)

Table 2-1. (*continued*)

File type	Features	Usage
JSON	• Record structure stored as key-value pair	• No support for block compression • Schema evolution easier than CSV or text file as metadata stored along with data

Why ORC is a preferred file format? ORC is a columnar storage format that supports optimal execution of a query through indexes which help in quick scanning of files. ORC supports indexes at the file level, stripe level, and row level. File and stripe indexes work similar to storage indexes from a relational perspective in that they help in quick scanning of data by narrowing down the scan surface area. They help in pruning out the stripes from scans during query execution.

Stripe indexes – An ORC file of a table is organized into stripes of default 64MB size. Stripe size can be configured at the table level. Each stripe implicitly indexes the column and holds meaningful details like min/max value or a dictionary for quick lookup. Some of the key ORC configuration parameters are listed below. Note that these parameters should be set at table level within TBLPROPERTIES clause.

1. orc.compress – Compression codec for ORC file

2. orc.compress.size – Size of a compression chunk

3. orc.create.index – whether or not the indexes should be created?

4. orc.stripe.size – Size of memory buffer (bytes) for writing

5. orc.row.index.stride – Rows between index entries

6. orc.bloom.filter.columns – BLOOM_FILTER stream created for each of the specified column

For more details on ORC parameter, you can refer to ORC Apache page – `https://orc.apache.org/docs/hive-config.html`.

For example, ORC file storage of CUSTOMER table (Figure 2-3)

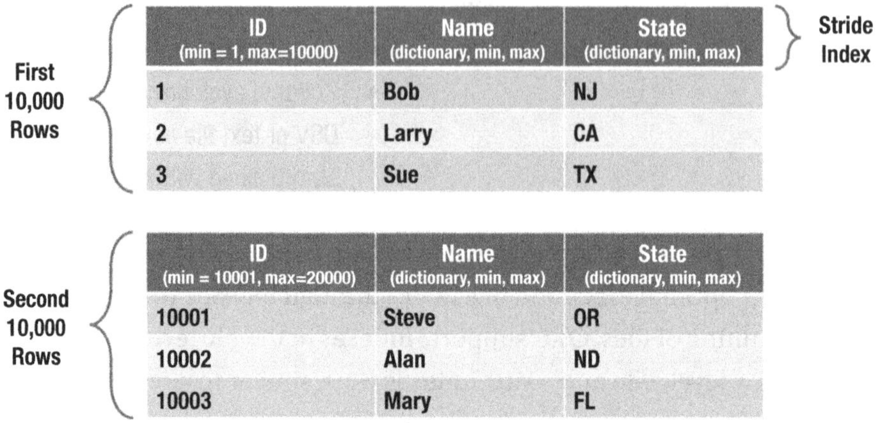

	ID (min = 1, max=10000)	Name (dictionary, min, max)	State (dictionary, min, max)	
First 10,000 Rows	1	Bob	NJ	Stride Index
	2	Larry	CA	
	3	Sue	TX	

	ID (min = 10001, max=20000)	Name (dictionary, min, max)	State (dictionary, min, max)
Second 10,000 Rows	10001	Steve	OR
	10002	Alan	ND
	10003	Mary	FL

Figure 2-3. *Stripes of CUSTOMER table*

A user issues the below query. The query filters the results on "state" column.

```
SELECT ID, NAME
FROM CUSTOMER
WHERE CUSTOMER.state = 'CA';
```

For CUSTOMERS table, the two stripes have 10,000 rows each. The number of rows in a particular stripe is configurable while creating a table. Each stripe contains inline indexes such as min, max, and lookup/dictionary for the data within that stripe. ORC's predicate pushdown will consult these inline indexes to identify if an entire block can be skipped all at once. The second stripe will be discarded because its index does not have the value "CA" in *state* column.

If a column is sorted, relevant records will get confined to one area on disk and the other pieces will be skipped very quickly. Skipping works for number types and for string types. In both instances, it's done by recording a min and max value inside the inline index and determining if the lookup value falls outside that range. Sorting can lead to very nice speedups, but there is a trade-off with the resources needed in order to facilitate the sorting during insertion.

ORC usage best practices

1. Hive queries must be analyzed to explore usage patterns and track down columns that frequently occur in predicates.

2. Hive tables must be timely analyzed to keep the statistics updated

3. Data should be distributed and sorted during ingestion. This will help in effective resource management during query processing.

4. If the filtering column in a query has high cardinality, then lower stripe size works well. If the cardinality is low, then a higher stripe size is preferred.

5. Starting hive 1.2, support for bloom filters was included to ORC semantics to provide granular filtering. It is used on sorted columns.

The ORC file format is supported by Hive, Pig, Apache Nifi, Pig, Spark, and Presto. On the adoption fronts, Facebook and Yahoo use ORC file storage format in production and have observed significant performance compared to other formats.

ETL vs. ELT

It would be an understatement that Extraction, Transformation, and Loading (ETL) protocol under-sufficed the data motion requirements for traditional data warehouses. It has been a standard de-facto process since the evolution of data movement strategies. However, with the next-gen data warehousing strategies and big data trends, the ETL approach tends to require tweaks.

Contrary to the traditional ETL approach, the data lake ingestion strategy adopts the ELT approach. With this approach, data gets loaded directly into the data lake after being collected. Transformation lies in the purview of the consumption or analytical layer (Figure 2-4).

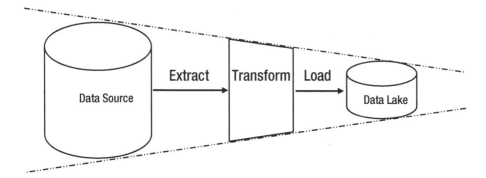

Figure 2-4. *Data agility is reduced in a typical ETL process*

Data lake is ideated to hold data from a variety of sources in its rawest form. A thin data scrubbing layer may optionally exist to clean raw data before it gets ingested into the data lake and consumed by analytical models. However, having a wide layer of data transformation is not recommended as it may restrict the surface area of data exploration, thereby narrowing down the data agility. Other rationale behind the ELT approach is the performance factor. Running transformation logic on huge volumes of data may foster a latency between the data source and data lake. The transformation layer can instead be flexed down to a curated layer to empower analytical models to retrofit the data stance. Figure 2-5 shows the data movement in an ELT model.

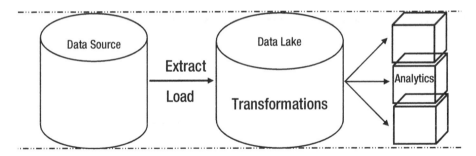

Figure 2-5. *Data agility remains intact in a typical ELT process*

Other factors that stand in support of ELT in data lakes are cost effectiveness and maintenance. Since the time data lake concept has caught all the eyes of data world, ELT has been the most trusted approach.

Big Data Integration with Data Lake

Data is a ubiquitous entity. Until the big data trend acquired the waves, it was the relational databases who held the system of records in a structured format. Although relational data store vendors are finding ways to address unstructured data, adoption is majorly driven by factors like cost, ease of processing, and use-cases.

Data lakes are designed to complement contemporary data warehousing systems by empowering analytical models to churn out the real value of "data" irrespective of its format. In this chapter, we will cover techniques and best practices of bringing structured as well as unstructured data into data lake. This section focuses on bringing structured data into data lake. We will walkthrough ingestion concepts, best practices, and tools and technologies used in the process.

Hadoop Distributed File System (HDFS)

Although we assume the readers of this book to be proficient with Hadoop concepts and HDFS, but to maintain the logical flow of concepts, let us get a high-level overview of HDFS.

The Hadoop Distributed File System constitutes a layer of abstraction on top of POSIX (or like) file system. During a write operation, a file is split into small blocks and apparently replicated across the cluster. The replication happens transparently within the cluster while the replicas cannot be distinctly accessed. Replication ensures fault tolerance and resiliency. Whenever a file gets processed in the cluster, all its replicas are processed in parallel; thus, bettering the computational performance and scalability.

The hdfs dfs command-line utility can be used to issue the file system commands in the Hortonworks distribution of Hadoop. In addition to this utility, you can also use Hadoop's web interface, WebHDFS REST API, or Hue to access the HDFS cluster.

```
hdfs dfs [GENERIC_OPTIONS] [COMMAND_OPTIONS]
```

1. Shell commands are similar to common Linux file system commands such as ls, mkdir, cat

2. Help commands –

    a. `$ hdfs dfs`

    b. `$ hdfs dfs -help`

    c. `$ hdfs dfs -usage <shell command>`

3. Directory commands like cd and pwd not supported in HDFS.

Copy files directly into HDFS

One of the simplest methods to bring data into Hadoop is to copy the files from local to HDFS. If there are bunch of csv spreadsheets, JSON, or raw text files on the local system, you can copy the files directly into HDFS using put command.

```
$ hdfs dfs mkdir /user/hdfs/sales_2017
```

Copies sales.csv from local to HDFS cluster

```
$ hdfs dfs -put sales_Q1.csv sales_2017
$ hdfs dfs -put sales_Q2.csv sales_2017
```

List the cluster files

```
$ hdfs dfs -ls /user/hdfs
```

Once the file is available in the Hadoop cluster, it can be consumed by Hadoop processing layers like hive data store, pig script, mapreduce custom programs or spark engine.

Batched data ingestion

In simple terms, batch is a frequency based incremental capture that kicks off as per the preset schedule. For most of the ETL frameworks, the implementation of the "extract" phase works on similar principles. Data collector fires a SELECT query (also known as *filter query*) on the source to pull incremental records or full extracts. Performance of the filter query determines how efficient a data collector is. The query-based approach to extract and load data is easy to implement with minimal failures.

From a relational data source, data can be extracted using the filter query by following either of the techniques listed below.

- Change Track flag – if each changed record (insert/ update/delete) on the source database can be flagged, the filter query can capture just the flagged records from the source table.

  - Primary key will be required to merge the changes at the target

  - If primary key exists on target

    - Delete the existing record

    - Insert the fresh record from changed data set

  - If primary key doesn't exist on target

    - Insert the record from changed data set

  - If the target table is modeled as type 2 SCD (slowly changed dimension), all changed records can be directly inserted to target table. A timestamp attribute or transaction id can be maintained on target to trace change history of a primary key.

- Incremental extraction – the filter query pulls the differential data based on a column that can help in identifying changes in the source table. It can be a timestamp attribute or even a serialized id column.

  - To apply the changes, primary key is a must

  - If PK exists, delete old and insert the new record

  - If PK doesn't exist, insert the new record

- Incremental extraction frequency – from the data consistency perspective, it is important to be aware when the source table is active for transactions and what is data change rate. If the change rate is high, incremental job should be periodically orchestrated.

- Full extraction – if the source database table is not very large and change frequency is low, target table can undergo full refresh every time the ETL runs. This ensures data consistency between source and target until source data gets modified. For source tables with master data and configuration data, full refresh approach can be followed.

Once captured from the source via filter query, the data extract needs to be staged on the edge node or ETL server, before its gets merged into Hadoop. This brings up the need for an additional storage system prior to the Hadoop cluster. The dual write approach adds to the latency and brings inconsistency in data lake.

Challenges and design considerations

An organizational data lake deals with all formats of data. Data, whether structured or unstructured, struggles with mutable data on Hadoop. Hadoop, being a distributed system relies on concurrency for functionality but dealing with mutability and concurrency could be meaty challenge. The ingestion framework must ensure that only one process updates the mutable object at a given time and avoids dirty read problems.

Other problems include datatype mismatch between source systems and hives, precision field handling, special character handling and efficient transfer of data with table size varying from Kilobyte (KB) to Terabyte (TB).

51

Design considerations

The issues discussed above are common in the target system, namely Hadoop data lake. The design considerations discussed in this section must be practiced on Hadoop objects.

1. Table partitioning – Splitting the data into small manageable chunks provides better control in terms of resource consumption and data analysis. Partitioning strategy should factor in the following parameters –

   a. Low-cardinality columns

   b. Frequently used in joins and query predicates

   c. Columns that can create interval based partitions

2. File storage format – ORC file storage format gives better compression compared to other file formats. In addition, it also stores index headers for optimized read access from files.

3. Full load or incremental - Full load integration should be practiced if change data capture is not possible. Data size and refresh frequency must be kept in mind while planning full load for objects.

4. Change merge strategy – If the target landing table is partitioned, then the changes can be tagged by table and the most recent partition. During the exchange partition process, the recent partition can be compared against the "change" data set to merge the changes. Figure 2-5 shows the process flow of strategy to merge change using exchange partition for partitioned hive tables.

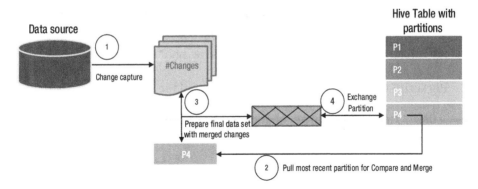

Figure 2-6. *Merge changes through exchange partition*

Let us consider a simple case of merging the changes using Piglatin. We have an interval partitioned hive table. The below code piece will show how to merge incremental changes from source data into a hive partition.

---sample data in a hive partition---

```
[bda@datalake sample-merge]$ cat hive_part4.txt
"20001""delhi"
"20002""mumbai"
"20003""bangalore"
"20004""chennai"
"20005""hyderabad"
"20006""pune"
"20007""kochin"
"20008""kolkata"
"20009""jaipur"
"20010""chandigarh"
```

Changes are captured via a change capture tool. The changed data set has a delimiter "ctrl A". Below is the change dataset that needs to be merged with most recent partition in hive table.

```
[bda@datalake sample-merge]$ cat change_dataset.txt
"I"^A"20089"^A"1"^A"2014-09-04 12:38:08.000"^A"20015"^A"noida"
```

53

```
"D"^A"20089"^A"2"^A"2014-09-04 12:38:08.000"^A"20003"^A\N
"I"^A"20089"^A"3"^A"2014-09-04 12:38:08.000"^A"20003"^A"bengaluru"
"D"^A"20089"^A"5"^A"2014-09-04 12:38:08.000"^A"20001"^A\N
```

Pig script to merge the changes with original file.

```
A = LOAD '/user/bda/merge_change/hive_p4_merged_set.txt'
using PigStorage('\u0001')
AS (
opcode:chararray
, seqno:chararray
, row_id:chararray
, commit_timestamp:chararray
, id:chararray
, place:chararray);
B = GROUP A BY id;
C = foreach B {
D = order A by seqno, row_id desc;
top = limit D 1;
generate flatten(top);
};
```

Check and verify the changes in main file. Note that [id = 20001] has been deleted, [id=20003] has been updated, and [id=20015] has been inserted.

```
[bda@datalake sample-merge]$ cat hive_p4_merged_set.txt
"20002""mumbai"
"20003""bengaluru"
"20004""chennai"
"20005""hyderabad"
"20006""pune"
"20007""kochin"
"20008""kolkata"
"20009""jaipur"
```

```
"20010""chandigarh"
"20015""noida"
```

Let's take another use case to demonstrate change-merge using Spark. We'll work with a main data set and changed data set. Master Data in Target Location

```
val main_data = spark.table(t.tablename).filter(cond) //filter
on the specific partition
```

We'll create two expressions using primary keys in the below fashion.

- Combining primary keys – pk1 AND pk2 ... pk_n

- Combining primary keys having null – pk1 is null AND pk2 is null ... pk_n is null

Below is the sample of Main Dataset A

P.K.	Name	VALUE	TIME_ID	DELETE_FLAG
1	Pranav	13341	10001	0
2	Shubham	18929	10002	0
3	Surya	12931	10003	0
4	Arun	12313	10004	0
5	Rita	12930	10005	0
6	Kiran	12301	10006	0
7	John	82910	10007	0
8	Niti	218930	10008	0
9	Sagar	82910	10009	0
10	Arjun	92901	10010	0

Below dataset represents the incremental changes captured via CDC mechanism

P.K.	Name	VALUE	TIME_ID	DELETE_FLAG
1	Pranav	13341	10020	1
2	Shubham	18929	10022	1
3	Surya	453202	10034	2
4	Tarun	489503	10098	0
5	Pranav	129789	10099	2

Here P.K. is the primary key column, TIME_ID is the defined value for timestamps and DELETE_FLAG is the value where 0 is termed as New Insert, 1 as Delete and 2 as an Update. The following spark code will merge the data and store it as a temporary view

```
main_data.as("m").join(broadcast(incr_data.as("k").
filter(cond)), expr(str1), "left_outer").filter(str2).
select("m.*").union(incr_data.filter("del_flag != 1")).createOr
ReplaceTempView(mergedTable)
```

Figure 2-7 shows the merge workflow process.

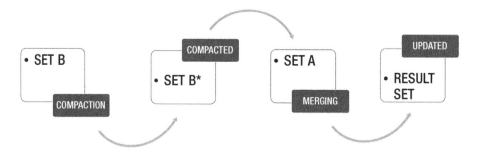

Figure 2-7. *Merge operation workflow process*

Below is the data set produced after merge.

P.K.	Name	VALUE	TIME_ID	DELETE_FLAG
1	Pranav	129789	10099	0
3	Surya	453202	10034	0
4	Arun	12313	10035	0
5	Rita	12930	10036	0
6	Kiran	12301	10006	0
7	John	82910	10007	0
8	Niti	218930	10008	0
9	Sagar	82910	10009	0
10	Arjun	92901	10010	0
11	Tarun	489503	10098	0

Commercial ETL tools

While the underlying principle of most of the 3rd party commercial ETL tools remain as discussed above, implementations can be different. For example, Informatica PowerCenter stores metadata in an Oracle database repository while Talend generates java code to do the job. Pentaho, on the other hand, provides a user-friendly interface.

Because data lake is a new opportunity, data integration software vendors have started complementing their ETL products with Hadoop centric capabilities. Modern-day ETL tools are flexible, platform agnostic, and capable of optimized extraction, through reusable code generation, and much more.

The 2017 Gartner magic quadrant (Figure 2-8) compares the data integration tools and positions Informatica as a leader.

Figure 2-8. *Gartner's magic quadrant for commercial data integration products.* `https://www.informatica.com/in/data-integration-magic-quadrant.html`

Real-time ingestion

A batched data ingestion technique is fool-proof as far as data sanity checks are concerned. However, it fails to paint the real-time picture of the business due to the lag associated with it. To enhance the business readiness of analytical frameworks, it is expedient to process a business transaction as soon as it occurs. In (near) real-time processing, changes are captured either at very low latency or in real-time. A log-based real-time processing exercise is known as change data capture.

Change data capture refers to the log mining process to capture only the changed data (insert, update, delete) from the data source transaction logs. A real-time or micro-batch CDC detects the change events by scanning the database logs as they occur. With minimal access to enterprise sources, CDC incurs no load on source tables; thereby minimizing latency and ensuring consistency between source and target systems.

So, why CDC? As we discussed in the last section, conventional ETL tools use SQL to extract and batch the incremental data. Query performance may be impacted due to continuous growth in source database's volume and its concurrent workload. In addition, the query incurs its portion of the workload on the source system.

Figure 2-9 shows a change data capture workflow between source and target systems.

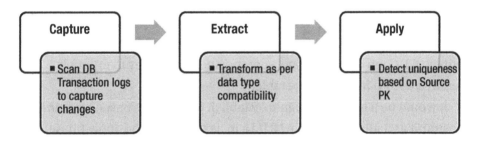

Figure 2-9. *Change Data capture workflow*

As part of business intelligence and data compliance initiatives, CDC helps in aligning with data-as-a-service principles by providing master data management capabilities and enabling quicker data quality checks.

Summing up the points, the CDC ingestion pipeline helps in –

- Eliminating the need to run SQL queries on source system. Incurs no load overhead on a transactional source system.

- Achieves near real-time replication between source and target

- Log mining helps in capturing granular data operations like truncates as well

CDC design considerations

To design a CDC ingestion pipeline, the source database must be enabled for logging. All relational databases follow a roll forward approach by persisting the changes in logs. Each and every event is persistently logged with a change id (or system change number) in a log and will never get purged. An Oracle database allows enabling supplement logging at the table level. Similarly, SQL Server allows logging at the database level. Without logging, transaction logs cannot be mined to capture the changes.

The tables at the source database must hold a *primary key* for replication. It helps the capture job in establishing uniqueness of a record in the changed data set. A source PK ensures the changes are applied to the correct record on target. If the source table doesn't have primary key defined, CDC job can designate a composite primary key to uniquely identify a record in the change table.

It would be a terrible design to establish uniqueness based on a *unique* constraint as it allows multiple NULLs in a column. In the apply phase, a change record with null identity will fail to pick a matching null record at the target.

Trigger based CDC –Another method of setting up change-data-capture is through triggers at the table level. A trigger helps in capturing row changes in a separate table synchronously, which apparently gets replicated to the target. Either the entire record is captured or just the changed attributes along with the primary key. The downside of this approach is that it induces overhead of one more transaction before the original transaction is deemed complete.

This method usually works in two scenarios –

- Logging not enabled on the source database

- Reading transaction logs is a tedious task due to its binary format

- T-logs not available for scanning due to software restriction or small retention time

So, should you always prefer CDC over batched ingestion? No. Real-time integration or CDC should be set up only when business demands it. It is a feature to be contemplated based on multiple factors like business's service-level agreement, change size, and target readiness.

Example of CDC pipeline: Databus, LinkedIn's open-source solution

Databus, a real-time change data capture system, was developed by LinkedIn in the year 2006. In 2013, LinkedIn released the open-source version of Databus. Since its development, Databus has been an essential component of the data processing framework at LinkedIn. Databus provides a real-time data replication mechanism with the ability to handle high throughput and latency in milliseconds. The Databus source code is available at its git repo at `https://github.com/linkedin/databus`.

Databus is a source agnostic framework that scales seamlessly to multiple consumers, while the transactional sources are still operational. The source code includes the adaptors for different relational sources like Oracle and MySQL. Figure 2-10 shows the working components of Databus.

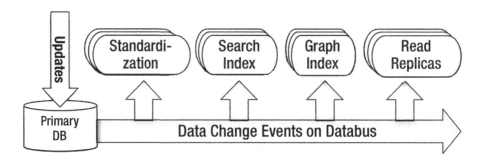

Figure 2-10. *Databus component diagram. Source: https://engineering.linkedin.com/data-replication/open-sourcing-databus-linkedins-low-latency-change-data-capture-system*

Databus works with these three most important pieces – relays, bootstrap, and client library. At a high level, the following list outlines the steps of Databus workflow.

- Relay is responsible for pulling the most recent committed transactions from the source

  - Relays are implemented through tungsten replicator

- Relay stores the changes in logs or cache in compressed format

- Consumer pulls the changes from relay

- Bootstrap component – a snapshot of data source on a temporary instance. It is consistent with the changes captured by Relay. (Refer to Figure 2-11)

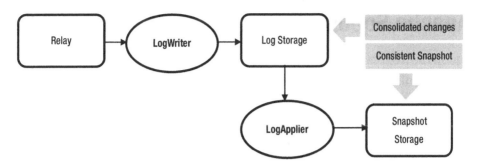

Figure 2-11. *Bootstrap component in Databus*

- If any consumer falls behind and can't find the changes in relay, bootstrap component transforms and packages the changes to the consumer

- A new consumer, with the help of client library, can apply all the changes from bootstrap component until a time. Client library will point the consumer to Relay to continue pulling most recent changes

Figure 2-12 branches out the benefits of LinkedIn's Databus solution.

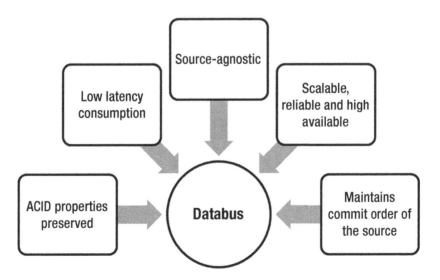

Figure 2-12. *Linkedin's Databus differentiators*

Apache Sqoop

Sqoop or "SQL to Hadoop" has been one of the top Apache projects that addresses the data integration requirements of Hadoop. It is a native component of the HDFS layer that allows bi-directional "batched" flow of data from the Hadoop distributed file system. Not just the users can automate data transfer between relational databases and Hadoop, but a reverse operation empowers enterprise data warehouses to augment their consumption layer with map-reduced data from data lake.

Apache Sqoop is available in two versions – sqoop 1 and sqoop 2.

Sqoop 1

The very first version of Sqoop was introduced in 2009. In August 2011, the project moved under Apache and quickly, Sqoop became one of the most sought-after ingestion tools.

Connectors are the motivation behind the working of Sqoop 1. The JDBC based connectors to different source systems are responsible for deriving metadata of source objects and data transfer. Let us list down the key highlights of Sqoop:

- Java based utility (web interface in Sqoop2) that spawns Map jobs from MapReduce engine to store data in HDFS

- Provides full extract as well as incremental import mode support

- Runs on HDFS cluster and can populate tables in Hive, HBase

- Can establish a data integration layer between NoSQL and HDFS

- Can be integrated with Oozie to schedule import/export tasks

- Supports connectors to multiple relational databases like Oracle, SQL Server, MySQL

Sqoop 2

Sqoop2 succeeded sqoop with a major focus on optimizing data transfer, easing of using extension framework, and ensuring security. Sqoop2 works on client-server architecture (service-based model) in which the server acts as the host for two critical components, the connectors and the jobs.

Sqoop2 features are as follows–

- Sqoop 2 can act as a generic data transfer service between any-to-any systems.

- Sqoop 2 comes with a web interface for better interactivity. Command line utility still works. Sqoop 2 web interface uses REST services running on sqoop server. It helps in easy integration with Oozie and other frameworks.

- Sqoop 2 employs both mapper and reducer jobs during data transfer activity. Mapper jobs extract the data, while the reducer operation transforms and loads the data into the target.

- Connectors will be setup on Sqoop 2 server which requires connection details to the source and targets. Role-based access to connection objects mitigates the risk of unauthorized access on source and target systems.

- The metadata repository stores connections and jobs.
  Connectors register metadata on the sqoop server to allow
  the connection to the source and the creation of jobs.

- The connector consists of partitioning API (create splits
  and enabled parallelism), Extract API (Mappers), and
  Loading API (Reducers)

Figure 2-13 differentiates Sqoop1 and Sqoop2 in terms of components
at sqoop processing layer.

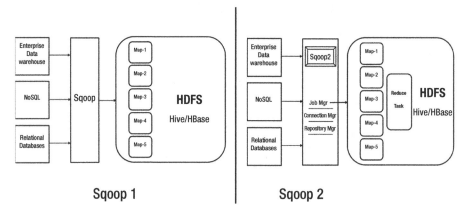

Figure 2-13. *Sqoop 1 vs Sqoop2*

How Sqoop works?

Sqoop adopts quite a simple approach to extract data from a relational
database. In a nutshell, it builds up an SQL query that runs at the source to
capture the source data, which later gets ingested into Hadoop. Let us look
at the internals of Sqoop.

Sqoop leverages mapper jobs of MapReduce processing layer in
Hadoop, to extract and ingest data into HDFS. By default, a sqoop job has
four mappers; this number is configurable though. Each of these mappers
is given a query to extract data from the source system. Query for a mapper

is build using a *split* rule. As per the split rule, the values of `--split-by` column must be equally distributed to each mapper. This implies that `--split-by` column should be a primary key. The entire range of PK is equally sliced for the mappers. Once the mapper jobs capture source data, either it is dumped in HDFS storage or loaded into hive tables.

Figure 2-14 demonstrates the primary key split mechanism.

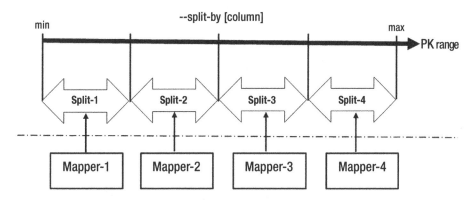

Figure 2-14. *Sqoop split mechanism*

Sqoop design considerations

Below are the key factors that can help in designing sqoop tasks effectively.

1.   Specify number of mappers in `--num-mappers [n]` argument

2.   Number of mappers

   a.   Note that mappers run in parallel within Hadoop, just like parallel queries

   b.   Large number of mappers might increase the load on source database. Decision should be taken based on size of the table and workload on the source database

    c.   Depends upon –

        i.   Handling of concurrent queries in the source database

        ii.   Varies by table, split configuration, and run time

3.   If the source table cannot be split on a column, use `--autoreset-to-one-mapper` argument to perform unsplit full extract using single mapper

4.   If the source table has all character columns with or without a defined primary key, we can have go with the below approaches –

    a.   Add surrogate key as primary key and use it for splits

    b.   Create manual data partitions and run multiple sqoop jobs with one mapper for each partition. This may cause data skewness and jobs will run for irregular durations depending upon the data volume per split

    c.   Character based key column can be used as `--split-by` column as usual, if the column has –

        i.   Unique values (or a partitioning key like location, gender)

        ii.   Integer values that can be implicitly type casted

5.   Sparse `split-by` column

    a.   Use `--boundary-query` to create splits

    b.   It works similar to retrieving split size from another lookup table

    c.   For text attributes, set

```
-Dorg.apache.sqoop.splitter.allow_text_
splitter=true
```

6.   Export data subsets

    a.   If only subset of columns is required from the source table, specify column list in `--columns` argument.

       i.   For example, --columns "orderId, product, sales"

    b.   If limited rows are required to be "sqooped", specify `--where` clause with the predicate clause.

       i.   For example, --where "sales > 1000"

    c.   If result of a structured query needs to be imported, use `--query` clause.

       i.   For example, --query 'select orderId, product, sales from orders where sales>1000'

7.   Good practice to stage data in a hive table using `--hive-import`

    a.   If table exists, data gets appended. Data can be overwritten using --hive-overwrite argument to indicate full refresh of the table

    b.   If table doesn't exist, it gets created with the data

    c.   Use `--hive-partition-key` and `--hive-partition-value` attributes to create partitions on a column key from the import

    d.   By default, data load is *append* in nature. Data load approach can be incremental by

    e.   Delimiters can be handled through either of the below ways –

        i.   Specify `--hive-drop-import-delims` to remove delimiters during import process

        ii.   Specify `--hive-delims-replacement` to replace delimiters with an alternate character

8. Connectivity – ensure source database connectivity from the sqoop nodes

    a.   Create and maintain a dedicated user at source with required access permissions

9. Always prefix table name with the schema name as `[schema].[table name]`

    a.   Supply table name in upper case

10. Connectors – common (JDBC) and direct (source specific)

    a.   Direct connector yields better performance

    b.   Use `--direct` mode argument with MySQL, PostgreSQL, and Oracle

11. Use `--batch argument` to batch insert statements during export

    a.   Uses JDBC batch API

    b.   Native properties of database (like locking, query size) apply

    c.   `Sqoop.export.records.per.statement`
        (10) – collates multiple rows in a single insert
        statement

    d.   `Sqoop.export.statements.per.transaction`
        (10) – number of inserts in a transaction

12.   Approaches to secure Sqoop jobs

    a.   For secure data transfer, use useSSL=true and
        requireSSL flags

    b.   Enable Kerberos authentication

13.   You can even create a Sqoop Spark job to enhance
       sqoop job performance

    a.   MapReduce engine might get slow with
        increased number of splits

    b.   No changes to the connectors. Enables
        pluggable processing engine

    c.   Spark job execution –

        i.   Data splits are converted to Resilient
           Distributed Dataset (RDD)

        ii. Extract API reads records, while Load API
           writes data

Native ingestion utilities

Ever since the Hadoop ecosystem reached a thoughtful stage, the tech stack has been able to provide extremely flexibility to implementers and practitioners. The big data ecosystem, in itself, comprises multiple pluggable components, which in turn, opens up a wide space for exploration and discovery. Ingestion patterns have evolved from tightly coupled utilities to standard and generic frameworks.

Many of the database software vendors who are planning their move to data lake, have developed home-grown utilities to facilitate transfer of its own data to Hadoop. What differentiates these native utilities from generic tools is the deep expertise in data placement strategy and the ability to capitalize on database architecture. In this section, we will cover utilities provided by the Oracle database and Greenplum to load data into HDFS.

Oracle copyToBDA

The copy to BDA utility helps in loading Oracle database tables to Hadoop by dumping the table data in Data Pump format and copying them into HDFS. The utility serves a full extract and load operation to Hadoop. If the data at the source changes, the utility must be rerun to refresh the data pump files. Once the data pump files are available in Hadoop, data can be accessed through Hive queries.

Note that the utility works on Oracle Big Data stack comprising Oracle Exadata and Oracle Big Data appliance, preferably connected via Infiniband network. It is licensed under Oracle Big Data SQL.

Under the hood, the utility uses ORACLE_DATAPUMP access driver and Hadoop client on Exadata to transfer the data. Figure 2-15 shows the workflow of the CopyToBDA utility.

Figure 2-15. *CopyToBDA utility workflow*

Additional notes –

1. Copy to BDA utility works well for static tables whose data change rate is not frequent. Reason being it doesn't allow the continuous refresh between source data and target.

2. If the table size is large, data can be dumped in multiple .dmp files

3. For a Hive external table to access the dump files and prepare the result set, specify appropriate SerDe, InputFormat and OutputFormat

   a. SERDE 'oracle.Hadoop.hive.datapump.DPSerDe'

   b. INPUTFORMAT 'oracle.Hadoop.hive.datapump.DPInputFormat'

   c. OUTPUTFORMAT 'org.apache.Hadoop.hive.ql.io.HiveIgnoreKeyTextOutputFormat'

Greenplum gphdfs utility

Greenplum offers the gphdfs protocol to enable batched data transfer operations between the Greenplum and Hadoop clusters. For Greenplum as a source, the utility has been a de-facto mechanism for data movement as it fully exploits the MPP capability of the database. On the target side, it can work with various flavors of Hadoop like Cloudera, Hortonworks, MapR, Pivotal HD, and Greenplum HD.

The gphdfs utility must be setup on all segment nodes of a Greenplum cluster. During a data transfer operation, all segments concurrently push the local copies of data splits to the Hadoop cluster. The number of segment nodes in the Greenplum cluster measure the degree of parallelism of data transfer. Data distribution on segments plays a key role in determining the effort at a segment level process. If a table is unevenly distributed on the cluster, the gphdfs processes will have an irregular split size, which will impact the performance of the data ingestion process.

The utility must be installed on each of the segment nodes. Installation steps are as follows:

1. Create repo file using

   ```
   wget -nv http://public-repo-1.hortonworks.com/HDP/
   centos7/2.x/updates/2.6.1.0/hdp.repo
   ```

2. Install the libraries using YUM

   ```
   yum install Hadoop Hadoop-hdfs Hadoop-libhdfs Hadoop-
   yarn Hadoop-mapreduce Hadoop-client openssl -y
   ```

3. Set the Hadoop configuration parameters

   a. `gpconfig -c gp_Hadoop_home -v " '/usr/ hdp/2.6.1.0-129'"`

   b. `gpconfig -c gp_Hadoop_target_version -v "'hdp2'"`

   c. Set java home and Hadoop home

Figure 2-16 demonstrates a schematic of a the gphdfs utility.

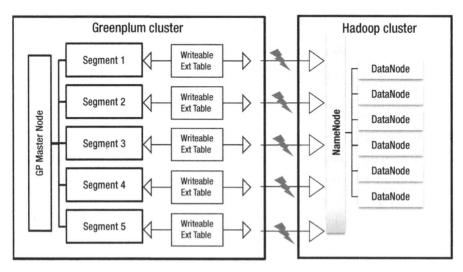

Figure 2-16. *How GPHDFS utility works*

Design considerations

1. JVM and gphdfs – The gphdfs protocol uses JVM on each segment host to access and write data into HDFS. While the writable external table is created on segment host and accessed via gphdfs, each segment instance initializes the JVM process with 1GB of memory.

   In case of high workloads during reading and writing multiple tables at the same time, JVM Heap memory issue might occur. You can decrease the value of the parameter `GP_JAVA_OPT` in `$GPHOME/lib/Hadoop/Hadoop_env.sh` from 1GB to 500MB.

2. Kerberos and gphdfs – The gphdfs protocol supports Kerberos authentication for Hadoop cluster. Kerberos authentication details are required to be updated in below files –

   - Yarn-site.xml

   - Core-site.xml

   - Hdfs-site.xml

   In addition, the /etc/krb5.conf must be present in the Greenplum cluster. In case you are facing GSSAPI errors while accessing HDFS, install the Java Cryptography extension (JCE) on Greenplum nodes ($JAVA_HOME/jre/lib/security).

3. Trigger gphdfs via ETL – The gphdfs utility can be embedded in Python script and fired through a standard ingestion tool like Informatica, Talend, Appworx, etc.

4. The LOCATION parameter of the writable external table must have either the Hadoop cluster name or HDFS namenode's hostname and port details.

5. Compression support – Use `compress` and `compression_type` arguments in writable external table to load data in compressed format into HDFS.

6. Custom loading framework is possible that loads group of tables (batch tables by schema or category) using python or any other scripting language

Data transfer from Greenplum to using gpfdist

In addition to gphdfs, the Greenplum utility gpfdist can be used to transfer the data from the Greenplum to HDFS.

The gpfdist utility offers parallel file operations in the Greenplum database. It can be used to move data from Greenplum segments to Hadoop clusters via edge node. You can create a writable external table in Greenplum using the below script.

```
CREATE WRITABLE EXTERNAL TABLE schemaname.tablename_ext
(LIKE schemaname.tablename)
LOCATION ('gpfdist://<edge_node_ip>:<port>/<location>')
FORMAT 'TEXT' (DELIMITER E'\x01' NULL '')
```

Once the table data gets exported to edge node, it needs to be pushed to the Hadoop cluster. There are two ways to copy this file to the Hadoop cluster –

1. Use Hadoop put command to copy file in HDFS

2. Secure copy (`scp`) the file to Hadoop name node

Ingest unstructured data into Hadoop

The technological and analytical advances sparked by machine textual analysis prompted many businesses to research applications, leading to the development of areas like sentiment analysis, speech mining, and predictive analytics. The emergence of Big Data in the late 2000s led to a heightened interest in the applications of unstructured data analytics in contemporary fields like natural language processing, and image or video analytics.

Unstructured data is information that either does not have a pre-defined data model or is not organized in a pre-defined manner. Unstructured information is typically text-heavy, but may contain data such as dates, numbers, and facts as well. This results in irregularities and ambiguities that make it difficult to understand using traditional programs as compared to data stored in fielded form in databases or annotated in documents.

Apache Flume

Apache Flume is a distributed system to capture and load large volumes of log data from different source systems to the data lake. Traditional solutions to copy a data set securely over network from one system to other, work only when data set is relatively small, easy and readily available. Given the challenges of a near real-time replication, batched loads, and volume, the urge to have a robust, flexible, and extensible tool cannot be ignored. Flume fits the bill appropriately as a reliable system that can transfer streaming events from different sources to HDFS.

Flume had its roots at Cloudera since 2011 and is packaged as a native component of Hadoop stack. It is used to collect and aggregate streaming data as events. Built upon a distributed pipeline architecture, the framework consists a Flume agent (or multiple independent federated agents) consisting of a channel that connects sources to sink. What flume guarantees is end-to-end reliability by enabling transactional exchange between agents and configurable data persistency characteristics of channels. The flume topology can be flexibly tweaked to optimize event volume and load balancing.

Figure 2-17 shows a simple data flow model from source to channel to sink via Flume. Flume agent is nothing but a JVM daemon process running on a machine.

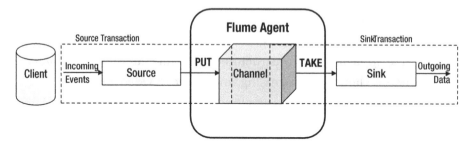

Figure 2-17. *Apache Flume architecture*

Components –

- A *flume event* is a byte size data object, along with optional headers as key-value pair of distinctive information, transporting through the agent.

- *Source* is a scalable component that accepts data from the data source and writes to the channel. It may, optionally, have an *interceptor* to modify events through tagging, filtering, or altering. Events pushed to the channel are PUT transactions.

- The *channel*, depending on its configuration, queues the flume events persistently as received. It helps in persisting the events and controls fluctuations in data loads.

- The *sink* pulls the data from channel and pushes to the target data store (could be HDFS or another flume agent). Events pulled by sink from the channel are TAKE transactions.

Data flow from source to sink is carried out using transactions which eliminates the risk of data loss in the pipeline. Flume works best for sources that generate streams of data at a steady rate. Source data can be *synchronous* like Avro, Thrift, spool directory, HTTP, Java message service, or *asynchronous* like SYSLOGTCP, SYSLOGUDP, NETCAT, or EXEC. For synchronous sources, client can handle failures, while for asynchronous, it cannot. Similarly, sinks can be HDFS, HBase (sync and async), Hive, logger, Avro, Thrift, File roll, morphlineSolr, ElasticSearch, Kafka, Kite, and more flume agents.

Tiered architecture for convergent flow of events

A tiered framework of multiple agents can be setup to enable convergent flow of events to multiple sinks. There can be multiple motivations behind the tiered approach. The primary motivation is to optimize the data volume distribution and insulate sinks from uneven data loads. Other reasons could be to relieve sources from holding large volumes of events for long time.

Loosely connected independent flume agents in the outermost tier (Tier-1) hold event streams from the sources. In the subsequent tier, sources consolidate the event streams received from preceding tier's sinks. The process of consolidation and aggregation continues until the last tier, before the sinks in the innermost tier route the events to HDFS. Agent count is maximum in the outermost tier while event volume is highest in the innermost tier.

Figure 2-18 shows three tiers, each containing multiple flume agents that read event streams from multiple web sources and transport data into HDFS cluster. Each sink pushes the event stream to the source of the agent in the successive tier. Tier-1 sources into Tier-2, which sources into Tier-3. This presents the scenario of *Consolidation.*

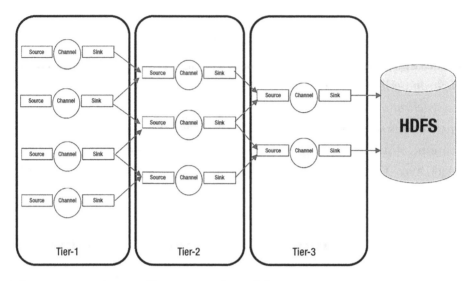

Figure 2-18. *Apace Flume tiered model*

A tiered architecture achieves load balancing and enables a distinguished layer between collector, storage, and aggregator agents.

Features and design considerations

1. Channel type – Flume has three built-in channels, namely, MEMORY, JDBC, and FILE.

   a. MEMORY – events are read from source to memory. Being a memory based operation, event ingestion is very fast. On contrary, since the changes captured are volatile in nature, incidents like agents crash or hardware issue can result into data loss. Business critical events are not a good choice but low category logs can be set of memory channel.

  i. You can set the event capacity using
      `agent.channels.c1.capacity`. Java
      heap space should also be increased in
      accordance with the capacity.

  ii. Use keep-alive to determine wait time
      for the process that writes event into the
      channel.

  iii. Low put and take transaction latencies
      but not a cost-effective solution for a large
      event

b. FILE – events are read from source and written to
    files on a filesystem. Though slow, it is considered
    as durable and reliable option amongst the three
    channels as it uses Write Ahead Log mechanism
    along with storage directory to track events in the
    channel. Set the `checkpointDir` and `dataDirs`
    attributes of the channel to set directories where
    events are to be held.

c. JDBC – events are read and stored in Derby
    database. Enables ACID support as well but acute
    adoption trends due to performance issues.

d. Kafka channel – events get stored in a Kafka
    topic in a cluster. This is one of the recent
    integrations that can be retrofitted into multiple
    scenarios:

  i. Flume source and sink available – event
      written to Kafka topic

  ii. Flume source – event captured in a Kafka
      topic. Integration with other applications
      is use-case driven.

      iii.   Flume sink – While Kafka captures the events from source systems, the sink helps in transporting events to HDFS, HBase, or Solr.

2. Channel capacity and transaction capacity – Channel capacity is the maximum number of events in a channel. Transaction capacity is the maximum number of events passed to a sink in single transaction. Attributes `capacity` and `transactionCapacity` are set for a channel.

    a.   Channel capacity must be large enough to queue many events. It depends on the size of an event, memory or disk size.

    b.   For MEMORY channel, channel capacity is limited by RAM size.

    c.   For FILE, channel capacity is limited by disk size.

    d.   Transaction capacity depends on batch size configured for the sinks

3. Event batch size – The transaction capacity or batch size is the maximum number of events that can be batched in a single transaction. It is set at the source and sink level.

    a.   Set at source – number of events in a batch written to channel

    b.   Set at sink – number of events captured by sink in single transaction before flush

    c.   Batch size `<<channel>>.batchSize` must be less than or equal to channel transaction capacity for proper resource management.

d. Larger the batch size at sink, faster the channels function to free up space for more events. For a file channel, post flush operation may be time consuming for fat batches.

e. Best practice to have transaction capacity that yields optimum performance. Not fixed formula but a gradual exercise.

f. If a batch fails in between, entire batch is replayed; which may cause duplicates at destination

4. Channel selector (*Replicator/Multiplexer*) – An event in flume, can either be replicated to all channels or conditional-copied to selected channels. For instance, if an event to be consumed by HDFS, Kafka, HBase, and Spark, channels can be marked as *replicator*. Replication is the default channel selector type. If an event needs to be routed to different channels based on a rule or context, selected channels can be marked as multiplexer. Selector applies before event stream reaches the channel.

```
agent.sources.example.selector.type = multiplexing
agent.sources.example.selector.mapping.healthy =
mychannel
agent.sources.example.selector.mapping.sick =
yourchannel
agent.sources.example.selector.default = mychannel
agent.sources.example.selector.header = someHeader
```

In case replicator and multiplexer do not suffice the requirements, custom replication strategy can also be developed.

5. Channel provisioning – if the channels are insufficiently provisioned in the topology, it will create a bottleneck in the event flow, in terms of event load per agent and resource utilization.

6. In a multi-hop flow or a tiered farm, keep note of the hops that an event makes before landing to destination. Note that the channels within the agents, at a given time, act as event buffers. In case of many hops, if any one agent goes faulty, the impact can get cascaded until source.

7. Flume follows extensible framework. Custom flume components are required to add their jars to FLUME_CLASSPATH in flume-env.sh file. Other way is the plugins.d directory under $FLUME_HOME path. If plugins follow the defined format, flume-ng process will read the compatible plugins from plugins.d directory.

8. Flume topology is highly dependent on use case. For a time-series evenly generating data, flume can work wonders. If source data pipeline is wrecked, flume is not a good choice as it might potentially break flume topology and cause prolonged outages. Frequent configuration changes to flume topology are not recommended.

9. Due to global spread out, time zones have become indispensable piece of data ingestion strategy. All timings and schedules must be normalized a single time zone UTC in its standard format.

Conclusion

In this chapter, we discussed different approaches to bring data into a the Hadoop data lake. The chapter kicks off with the principles of ingestion framework and a quick brush up on basic ETL and ELT concepts. We discussed batched ingestion concepts and its design considerations. Under real-time processing, we explored how change data capture works and what its key drivers are in real-world scenarios. Key takeaways from this chapter would be two apache foundation products: sqoop and flume. Both have proved useful in integrating structured and unstructured data in data lake ecosystems.

In the next chapter, we'll cover data streaming strategies, focusing majorly on Kafka.

CHAPTER 3

Capture Streaming Data with Change-Data-Capture

"Data will talk to you, if you're willing to listen to it"

—*Jim Bergeson, President and CEO of Bridgz Marketing Group*

It would be fruitless to design a data lake without factoring in the velocity of data flowing from data sources. Streaming data sources are becoming increasingly critical from a real-time or "freshness" perspective. The era where the internet of things and mobile trends carries as much prevalence as human rights, demands a system that not only matches the pace of data flow, but also puts it into action. Data lake beneficiaries and analytics consumers face the tough task of ingesting the continuous motions of data.

With challenge comes the techniques to overcome it. As the complexity of data and its producer sources grow at a steady rate, the data engineering community has evolved to a level where data streams can be ingested successfully. As mentioned earlier, the big data ecosystem is highly flexible and solution trends are under continuous evolution. This chapter will uncover the principles of change, data capture, and event

© Saurabh Gupta, Venkata Giri 2018
S. Gupta and V. Giri, *Practical Enterprise Data Lake Insights*,
https://doi.org/10.1007/978-1-4842-3522-5_3

stream processing using Kafka. We'll try to understand the categories of change data capture approach while ingesting data into data lake. In the second half of the chapter, we see how to publish events using Kafka. At a broader level, the chapter is structured as below –

- Change Data Capture
  - What and Why?
  - Strategies and tradeoffs
  - Retention and Replay
- CDC Tools
  - Operation and Challenges
- Downstream Propagation
  - Centralization of Change Data
  - Analysis: Centralized Data Store
  - Metadata: Data about Data
  - Data Formats
  - Consumption and Checkpointing
  - Merging and Consolidation
  - Data Quality and challenges
  - Publishing to Kafka

Change Data Capture Concepts

Change data capture (abbreviated as CDC) is a key process within any company's overall data management and consolidation strategy. A simple definition of change data capture is, the extraction of a committed state changes for a piece of data from its host data store. Data stores can be traditional enterpsie databases like Oracle, MySql, or NoSQL stores like Mongodb or Cassandra. In reality any proprietary format, as rudimentary as a pipe-delimited data store, can potentially be a source of truth for change data capture. The term "committed" is significant in this context because it (or the lack of) could lead to data inconsistency, such as capturing the changed data before the final state transition (specifically rollback in this case) results in data mismatch between the source systems and the downstream applications.

Most of the time, an input record or byte of data collected by a system doesn't fulfill its existential promise just by serving the read traffic on that data. It needs to flow to other downstream systems to serve a larger big-picture purpose. In fact, even when the sole purpose of existence for a piece of data is to serve the read clients querying the original data store, there could still be a valid change data capture requirement for it, for validation and tracking/compliance purposes.

Strategies for Data Capture

There are different ways of capturing changes to data in a data store. The best approach depends on various parameters like the nature of data and the rate of flow of input data. The capture strategy could span across multiple phases including synchronous data capture and aggregation of capture data from synchronous capture.

The optimal strategy also depends on the requirements driving the data capture itself that boils down to a few questions listed below.

- What's the purpose/goal of capturing change data?

- Is it for tracking/compliance purposes (or) aggregation/ analytic reporting?

- What are the SLAs for data availability for the destinations that consume from this data capture?

All of the above factors are critical in determining the strategy for capturing and propagating the data to the next step of the data flow process/pipeline.

Following are some of the attributes that would help drive the change data capture strategy.

1. **Nature of Input Data**

    - Data Type - integer, float, string, binary, etc.

    - Size - average record size, average field size

    - Nature of change - percentage of inserts, updates, deletes

    - Rate of change - frequency of each of the types of changes (insert, update, etc.)

    Following are a few examples of the possible impact of the above attributes.

    - Free text data types like string could topple size and capacity estimates as they could potentially accommodate wide size ranges of data. This becomes even more unpredictable with large text data types like CLOBs. Therefore, a synchronous capture strategy (ex: trigger) on widely varying size ranges could end up derailing latency numbers for the write clients.

    - Similar size issues could surface based on row sizes. Some downstream systems (like KKafka) may have message size limits; this could potentially alter our data capture strategy depending on how we want to handle this scenario.

    - Update, inserts, and deletes could result in vastly different message sizes in cases where we capture "before" and "after" images.

2. Drivers of CDC

   - Purpose - Tracking/Compliance, Bulk Aggregation, Analytics etc.

   - SLA - (Near) real-time, Within the Hour, Within the day etc.

   - Following are a few examples of how the above attributes could drive CDC decisions.

      a. Tracking/Compliance systems typically need less information to be logged when compared to, say, data aggregation purpose. Therefore, synchronous capture may be a feasible strategy because of lower data processing latencies involved.

      b. SLA is obviously a key driver because it dictates how close the data capture should be to actual data commit time.

Retention and Replay

Data Retention for extracted data typically adds a lot of value in a CDC pipeline. Its main purpose is to facilitate re-use of extracted data in case of failures and delays, and thereby avoid expensive re-extraction of data in terms of resources and time.

   Retention is leveraged due to the typically high cost of data extraction. Depending on the mode of data extraction, synchronous or asynchronous, the source system incurs extra "cost" as significant resources are used to extract data. In case of any failure, re-doing the entire process would be expensive and/or infeasible.

Data retention allows for replay of data, namely, reuse of extracted data for downstream processes. Different downstream processes can replay from different points in the past with basic checkpointing mechanisms in place. Checkpointing, here, means saving the point until which a given consumer has successfully consumed data.

Retention Period

The retention period should be determined from the ETL pipeline requirements. Typically, there would be quite a few factors at play, but arriving at a retention number should be straightforward once all the dependencies are resolved.

For example, in a simple ETL flow data is extracted from source database, asynchronously through queries, and saved in a delimited format within files. Later, a downstream process consumes these files at an hourly frequency and merges this data into hadoop. In this simple use case, the key factors that determine data retention are –

1) Source data volume/transaction rate

2) Source data extraction frequency

3) Organic/expected growth in data volume

4) Hadoop merge frequency

5) Analysis of storage cost vs data extraction cost

There could be other factors like host capacity on which the data extraction process runs, but let's assume other factors are easily addressable, for simplicity's sake.

The impact of factors #1 and #2 is straightforward. These numbers determine how much data will be produced because of data extraction, every time the extraction process runs. That will give us an estimate of how much storage we would need to meet the data requirements.

Factor #3 is essential to determine the storage requirements for future growth. For example, if the company is rolling out a new feature in the next few months that could drastically increase data and transaction volumes, this would obviously need to be factored into the retention planning. Also, there should always be some buffer room for unexpected spikes or growth in data volume.

Factor #4 is important because it provides the minimum retention required for the extraction flow to be practically useful for the downstream process. For example, if the Hadoop merge process runs daily and change data extraction happens hourly, then the extracted data from CDC needs to be stored for at least a day, for the Hadoop merge process to be able to consume without gaps in data. Further, this would provide an estimate of possible replay requests from the downstream process, if the merge fails.

Factor #5 is to determine if it's a trade-off between storing lot of data for a long period and re-extracting it as required (for failure and replay scenarios). Most of the time, extra retention for extracted data would be the better option compared to re-extraction of data from the source.

This approach can be extended to any data flow pipeline. There could be other factors relevant to specific use cases like host capacity and network bandwidth. They have to be factored in, as applicable, in determining the data retention policy.

Types of CDC

Let us look at the different modes of change data capture – incremental, bulk, and hybrid.

Incremental

The mode works on capturing changes at commit time or close to that. The primary intention here is to extract data when change happens so that the capture load becomes manageable. Incremental mode can be synchronous or asynchronous.

Synchronous capture captures data changes as they happen at the source. While this approach has the advantage of the lowest possible capture load and delay time, it could add to the latency of applications or systems that persist data to the source data store. Clients that write to the source data store will have to sign off the extra latency of capture time, for this approach to be feasible. The extra latency incurred by clients also will change with input data volumes and commit chunk sizes. All these factors must be considered for this approach to be successful and scale over time.

Asynchronous capture captures the changes from the transaction logs after the changes have been committed in the source database. It has no effect on the transaction as the changes are read form the logs after the transaction is committed. Supplement logging must be enabled at the appropriate level to achieve asynchronous capture.

Bulk

Bulk capture captures data in chunks of extraction spread over time, like hourly data extracts. This bulk extraction could be on the source data store itself or on aggregation/ETL preparation on data extracted through synchronous capture.

If bulk extraction happens on the source data store, there will be an extra load of data extraction when the extraction process is run. The impact of this strategy on the source data store will depend on the resource usage level on the data store and related hardware. High impact will translate to latency impact on clients as well, albeit in an indirect way.

Hybrid

You could leverage a mix of both incremental and bulk strategies for change data capture. You could go with multiple layers of CDC that include incremental and bulk change capture. While this approach will bring with it the disadvantage of increased latency on the data store clients, it also has the advantage of incremental data capture discussed earlier. In addition, the approach can potentially optimize change data that is propagated to downstream or intermediate centralized systems.

For example, if your incremental capture is synchronous and extracts all changes to a given database record, this could result in significant amounts of change data for downstream systems to consume because of potentially numerous versions of a given record. This will result in an increased data footprint in multiple downstream destinations.

While a lot of data is good in general, it could also lead to data overload if your downstream is just interested in merging all changes and maintaining a consolidated copy of the source systems. In this use case, a second layer of bulk capture can consolidate and de-duplicate over a given time period, say, hourly, and propagate it to the next downstream layer, if SLA increase is agreeable

In general, the change data capture approach needs to consider various parameters within and around the data pipeline. The decision here will have to balance pros and cons, to arrive at an optimal strategy. Let's look at some key trade-offs in the CDC strategies we discussed.

CDC – Trade-offs

As the case with all theoretical principles, there will always be tradeoffs to be factored in during practical implementation. The best choice may not be an ideal one; but it's the most favorable and desired one that travels through. We did hint at a few trade-offs in our general discussion of strategies and means to coming up with a good change data capture process. Let's look at them in detail.

Synchronous-incremental capture offers the fastest and most true-to-the-original means of extracting system changes. Truthfulness to the original, in this context, means minimal or no chance for corruption because we extract changes as part of the persisting process itself. The trade-off is the latency; the data writer process takes the hit in terms of latency. This can be mitigated to some extent by making the incremental extraction slightly delayed or asynchronous, that is, we still extract data incrementally within a short time of persistence (few seconds or minutes), but it's still isolated from the original writer. However, to be able to do this, we'll need a place to go to for the original transaction details.

With relational databases like Oracle and MySQL, redo logs and bin logs respectively have the running trail of transaction activity on the database. Change capture process could, in theory, extract data asynchronously from these logs with minimal or no impact on the online database. Of course, in some cases log format could be proprietary, and extraction of data from logs may require some additional tools. We will talk more about these tools in a later section. But this asynchronous approach is a good trade-off because it still offers extraction capabilities close to source data persistence time, without significantly impacting the performance of source data store.

This trade-off could also be hybrid where synchronous extraction can be restricted to bare minimum data volume, namely, to just the raw, simple data, which can be later, asynchronously processed to meet the full data capture requirements.

For instance, if the data capture requirement is to send point-in-time representation of a data record across three customer related tables, we could synchronously extract individual records (ex: through triggers) and later asynchronously join these records to provide a full version of de-normalized records to the data capture process. With bulk capture, the frequency of running the bulk process could significantly contribute to the efficiency of the process.

For example, the amount of data that will get processed within a 1-hour capture interval could be manageable 80% of the time, but if it still is too big 20% of the time, it could significantly impact the reliability of the overall ETL pipeline. If this 20% use case causes significant delays in data availability through the CDC pipeline, increasing the frequency of data capture (to say, once every 30-minutes) could result in a much smoother CDC pipeline.

CDC Tools

Change data can be extracted through off-the-shelf tools as well as through simple query-based extraction. These queries could be in SQL or API calls based on the type of source database. Run the queries on the "delta" column (a timestamp column that tracks changes) periodically, at the desired frequency, to extract data and write it to a location (shared storage like netapp, filer, another data store, etc.) that would feed downstream processes. The query frequency has to be set based on other factors like desired SLA, load on the source database, peak volume of incremental data.

There are a variety of tools available, free and licensed, for incremental data extraction from RDBMS and NoSQL databases. Here are some of the commonly used ones.

- Database triggers can provide good data extraction mechanism for relational databases like Oracle or MySQL. However, as we discussed earlier, triggers are synchronous and could result in increased latency for data writers. In general, since triggers would have performance implications on the source database, the logic within triggers should be kept very lightweight. If required, an extra asynchronous layer of data extraction can be applied on top of data extracted through trigger to accomplish further aggregation needs on data, within the ETL pipeline.

- Oracle GoldenGate Adapters (Big data and Java adapters) for Oracle and MySQL. However, these tools need extra licensing, on top of Oracle GoldenGate license.

- Open source tools like Open Replicator, Tungsten Replicator for MySQL.

- Views for Couchbase

Challenges

The key challenge in a CDC system is to scale it appropriately for current and future needs. Scaling strategies depend on current data volume and speculative growth combined with data retention. These factors could vary for different use cases, as we discussed under the data retention section. Failure to appropriately scale and plan for expansion will result in huge operational overhead for the system as well as significant new effort for enhancing the system to meet the new state of things.

Ease of operation is another key factor that typically gets ignored during development phase. A system that is operationally heavy with low availability index on a day-to-day basis, is practically a non-functional system.

Downstream Propagation

Data captured from source data stores will need to flow to different destinations depending on the goal of overall ETL (extract, transform, load) process.

For example, if the goal is to get all data from various source data stores into a bulk data store like hadoop, you'll need a strategy to capture (incremental or bulk) data at the sources and direct them to hadoop. Depending on the granularity of data required on hadoop, there could be intermediate stages in this end-to-end data pipeline.

Further, there could be other downstream consumers that need the same data. In this case, the optimal approach would be to leverage the captured data for all downstream consumers, without having to setup multiple capture processes for different destination flows. Let's dive into this scenario further, through a simple, practical use case.

Use Case

Company A is a consumer web company that stores online data, from user-facing applications, into MySQL databases.

Downstream #1:

Source MySQL data should flow into Hadoop where the data is combined with data from third-party data from online user activity tracking company.

Downstream #2:

Online applications need to display user data from MySQL databases. For scaling purposes, applications use caching layer to speed up data access. So, the data from MySQL databases also needs to flow to these application caches.

Let's say, there's a data capture process that does bulk extraction every hour by running queries on the source databases. These queries are run on read-only slaves of the online databases.

Assumptions – Granularity of data in Hadoop is at record levels within MySQL database i.e, the records from source databases need to be propagated as they are into hadoop. Granularity of data in cache is at record levels as well.

Now, if the overall ETL strategy is to extract data, process it in chunks, and then dump into the destination environment Hadoop (or) cache. We'll need two pipelines that extract, process, and load data into destination.

But since we extract the source data in its original granularity in at least one of the flows, this extracted data can clearly be re-used across both the pipelines. To do this, we can extract once and reuse the captured data to direct it into multiple downstream destinations. In other words, having a central repository of data obtained through change data capture would save a lot of time and resources, and benefit our ETL strategy immensely.

Centralization of Change Data

Centralization of data enables a one-to-many consumption pattern for source data. It avoids duplication of data flows and facilitates easy consumption of data by downstream systems, for example, publisher-subscriber model. In addition, for a slow moving external data, centralization acts as a "staging" layer, that holds all the changes and efforts to search through the changes are eliminated. This layer yields much more benefits than its operational overheads. Figure 3-1 shows how multiple consumers can be served through single capture from source data systems.

Further, even if downstream destinations need data at different granularities, a centralized store maintained at the lowest required granularity could feed all or some of them. For example, a centralized data store with record level granularity could feed a full-aggregated-data downstream data store as well as a monthly roll-up downstream data store that combines monthly rollups of the same data with data from other sources.

Centralized data stores offer many other advantages as listed below –

- Better manageability because of fewer locations from which data fans out, compared to the case of several independent ETL pipelines.

- Control over downstream consumers through checkpointing and watermarks.

- Replay capabilities, without having to re-extract from source; if CDC runs on live/online systems, this also means reduced impact on live systems.

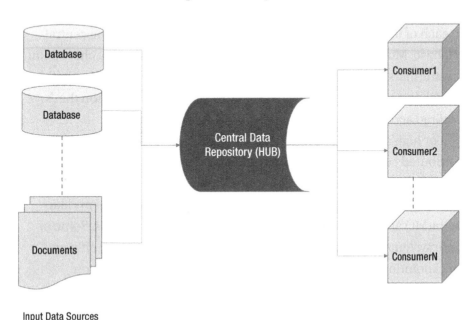

Input Data Sources

Figure 3-1. *Centralized data hub helps in serving multiple consumers through a single capture*

Analyzing a Centralized Data Store

For low transaction-volume use cases, the centralized store could be an aggregation of flat files. Data extracted from source data stores could be dumped into flat files that can be retained for a period that serves the pipeline requirement of consumption and replays across multiple consumers.

For slightly complex use cases that require better organization, query capabilities, and consumer checkpointing the centralized store can be a bigger data repository modeled after one of the source data stores.

For example, if you're building an ETL pipeline for MySQL source databases that does not have high data volume (few GBs incremental data per day, at most), you could persist all the incremental data from various sources in a central MySQL database. You could run incremental SQLs on each of the source databases and channel the output to one central database. Multiple consumers could consume data through a checkpoint or watermark mechanism built on top of this data. Data retention in this central data repository will be driven by the maximum retention that's required across all the consumers.

For high and very high-volume scenarios, a more scalable and distributed centralized data store would be the way to go. Publisher-Subscriber mechanism of consumption can be used to facilitate data consumption for multiple clients. Kafka and other queue systems like RabbitMQ and ActiveMQ can be employed to accomplish this. Some of these systems provide distributed storage of centralized data, efficient checkpointing for multiple consumers, and parallel consumption capabilities.

Metadata: Data about Data

For any data pipeline to be fruitful, actual data that flows within it should be accompanied by metadata that defines the data. Metadata is the key to effective data governance. Metadata in this context is the data that defines the structure and attributes of data. This could mean data types, data privacy attributes, scale, and precision. In general, quality of data is directly proportional to the amount and depth of metadata provided. Without metadata, consumers will have to depend on other sources and mechanisms.

The metadata management holds a critical piece from a compliance and security standpoint as well. The General Data Protection Regulation targets control over personal data for companies worldwide (who have EU citizens) to strengthen data protection. Figure 3-2 shows the areas where metadata can help in adhering to governance guidelines.

Figure 3-2. *Metadata management plays critical role in data regulation*

Metadata can be synchronous or asynchronous to the data itself. In case of asynchronous delivery, care should be taken to provide metadata as close as possible to the time of availability of the source data; otherwise it could break some downstream systems. But asynchronous mode prevents tight coupling of metadata with data especially in cases where metadata is not available at the time of data extraction and must be fetched separately.

When a CDC tool is used for data extraction, if the tool doesn't provide complete metadata at the time of data extraction, there will need to be a separate process that extracts metadata from the source system(s).

The nature of metadata extraction (synchronous or not) should be determined based on the criticality of metadata. If downstream systems are okay with occasional increased latency for metadata availability (at least some of it), asynchronous mode would be the better choice since it reduces dependencies and thereby the chances of overall failure. However, clear SLAs need to be defined for availability of metadata. The general requirements of scalability and high availability are applicable to metadata as well. Even if one downstream consumer fails because of non-availability of metadata along with data, overall data pipeline becomes unavailable.

Let's look at some features of metadata that typically would be required along with data.

Structure of Data

Data type precision (especially for numeric types, such as dates) is a key requirement for consumers, to preserve data fidelity. For example, scale and precision information for numeric data types is very important in financial aggregations and calculations on data.

Privacy/Sensitivity Information

Many data stores that store any member data would have few sensitive fields that need to be masked/scrubbed when exposed to few downstream systems, for example, SSN, credit card, and revenue info. These fields should be tagged with required metadata for appropriate downstream handling.

Special Fields

Another common use case is contextually relevant fields that are used across a set of consumers.

For example, the last modified timestamp is a common value that would be used for incremental data consumption and merges, within downstream systems. In these cases, metadata should identify the attribute that has this value. Another example is an attribute that identifies if a record is active or not.

Data Formats

Data pipelines are practical only when they contain the data and relevant metadata that consumers can use to make sense of the data.

As we discussed earlier, there are different ways to propagate metadata within the pipeline, along with the data. Whatever way we choose, it brings up a new problem of data size: the practical aspect of bundling metadata information with the data. The more metadata, the bigger the overall data set size for the records, and the slower the throughput (all other things being the same).

This is where the data format we choose becomes paramount. As always, there are trade-offs in the process and practicality has to drive the decision of choice. Let's look at some practical choices.

Delimited Format

The delimited format is very commonly used: comma/space delimited or some custom sequence delimited text. This choice is a very simple and efficient one w.r.t data size; there's very minimal storage overhead w.r.t actual data size (just the extra space for delimiters). However, delimited format prints just the data and metadata needs to flow through a different pipeline.

The advantage of isolation between data and metadata and no-dependency could be a disadvantage in the following scenarios:

1) Downstream consumers need the metadata to make any sense of the data.

   a) In this case, consumers will be blocked until metadata arrives.

   b) However, if the metadata is fairly static, consumers could cache metadata locally and use it to process data. When the metadata changes, there could still be some delay in processing; but if this change is very infrequent, this may be an ok trade-off for the benefit of compact and high-throughput delimited data

2) Metadata changes frequently.

a) When metadata changes frequently, consumers will have to look up metadata very frequently, from a different place (that stores the metadata). This will increase the pipeline latency and possible points of failure.

Avro File Format

Avro is a very commonly used format that embeds metadata along with the data itself. This allows for synchronous metadata propagation and validation as well. When an Avro schema is defined for data, the data that comes in against the schema is validated at run time.

For example, if a field is defined as number, any attempt to encode a string into that field, at run time, will be rejected. This strict validation of data type and structure goes a long way in terms of data standardization and provides great reliability for the pipeline when there are multiple consumers. Lack of validation could result in frequent failures due to type mismatches and improper communication between teams that manage data source and downstream destinations.

Avro encoded storage format is reasonably compact as well especially with binary encoding. With Avro, JSON encoding can be used as well.

While Avro is an efficient encoding format, how much metadata to encode synchronously is still a tradeoff that needs to be evaluated practically. For example, data type information is required for defining the Avro structure and for validation as well. However, extra metadata, like sensitivity information, would have to be provided explicitly. The mode of propagation for this extra metadata, synchronous or asynchronous, still has to be driven by practicality considerations.

Consumption and Checkpointing

Once a centralized data store is built to host data from source data stores, there needs to be well-defined processes to support multiple consumers.

Simple Checkpoint Mechanism

A simple organizational mechanism is to maintain a table of consumers with their current checkpoint information. Checkpoint itself could be a timestamp value that denotes the last record timestamp value a given consumer has extracted. It could also be an increasing number that denotes the last consumption point. Databases like Oracle provide a number that could potentially be used downstream for checkpointing. A similar number can also be generated for other data stores during extraction (CDC) process.

Parallelism

The consumption and checkpointing process becomes complicated when consumers need parallelism. For example, if one consumer can run multiple consumption threads in parallel to consume the same data, managing checkpoints can be a lot more difficult.

There are many publisher-subscriber model stores like Kafka that support parallel consumption, across distributed nodes. They have efficient ways to deal with data scaling across multiple nodes and facilitate multiple consumers and checkpoints as well.

Merging and Consolidation

Once data is available in a central hub or repository from various data sources, the next step in the data pipeline would be to build a sub-pipeline to consolidate this data with destination data store(s).

The exact nature of consolidation would depend on the nature of incoming data and the desired state of the destination data set/store. For example, if the data comes in through incremental dumps and the requirement is to maintain a consolidated, point-in-time version of the entire data set, the incremental incoming data sets need to be merged with destination data sets to arrive at the desired state of data at the destination.

If the incoming data set is a full data set every time, that is, if full dump of source data arrives every time, we can delete and replace the destination data set every time. We can move the target to a different place or data store if we want to retain old versions as well. This approach generally works for small to medium data sets because generating a full data dump of large source data sets requires a lot of resources; further, delays in processing could pile up to impractical data volumes to be processed. Figure 3-3 depicts the merging and consolidation process from centralized data hub to consumer to consumer processes.

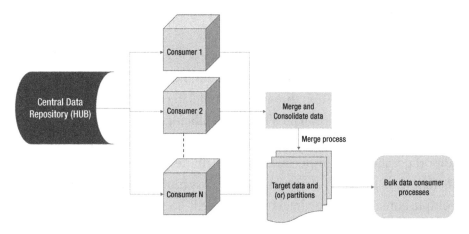

Figure 3-3. *Merging and consolidation workflow*

Design Considerations for Merge and Consolidate

Merging and consolidation are key phases of building the destination data set or data store that in turn can serve various end client use cases. These use cases can be for on-the-fly analytics or bulk processing. Typically, this destination data set becomes the source of critical reports that provide significant value to end users or internal customers.

For example, destination Hadoop data store can serve internal business analytics or user analytics on a web site. Any error in consolidation will have repercussive effects on the use cases that use this data. So, it is very important to make sure the consolidation process is robust and scalable. Let us list some key design considerations.

1) Simplicity of consolidation/merge logic - This goes a long way in making sure the destination data set is reliable and easy to debug, in case of issues. Simplicity, here, means less scope for data corruption or loss, and a clean and simple set of rules to consolidate data. For example, if you are required to resolve conflicts as part of a data merge, make sure the conflict resolution process is plain and simple (as in case of last-writer-wins i.e., record with latest timestamp wins). Avoid providing too many options as much as possible.

2) Repeatability of apply/merge process – Things could fail within a data pipeline from time to time, for various reasons. Apply process should allow for data replay i.e., running the apply process for same data (all or some of already applied data) multiple times should not result in data corruption or loss at the destination. This feature also allows for isolation of tasks within the data pipeline. Apply process would be able to run independent of other upstream processes.

109

3) Manageability of Target Data – Data in many enterprises typically flows from transactional data stores to bulk data stores. This is because the destination data stores are meant to serve the purpose of facilitating analytics and bulk aggregation on data from disparate sources. Target/destination data store size would be typically much larger than any individual source data store. More importantly, destination data stores allow for bulk operations on large volumes of data.

Most of the bulk operations benefit a lot from parallel processing of underlying data. If underlying data has inherent parallelism built within it, this factor would optimize the data processing significantly. Partitioning the data set appropriately could provide considerable performance benefits in bulk processing of destination data. Partitioning should be done based on attributes that participate in bulk operations.

For example, if there's an analytic query use case of "number of comments per user" that is generated on a bulk social network data set (of comments), partitioning the data set by user ID ranges will allow for parallel processing of underlying aggregate computation.

Data Quality

The quality of data that flows within a data pipeline is as important as the functionality of the pipeline. If the data that flows within the pipeline is not a valid representation of the source data set(s), the pipeline doesn't serve any real purpose. It's very important to incorporate data quality checks within different phases of the pipeline. These checks should verify the correctness of data at every phase of the pipeline. There should be clear isolation between checks at different parts of the pipeline. The checks include checks like row count, structure, and data type validation.

For example, quality of data should be verified w.r.t source data store after extraction; likewise, data in centralized data store should be verified against consolidation/merge output. If there are other intermediate stages within the pipeline, similar quality checks should be incorporated to compare input and output of each phase. This demarcation allows for clear and practical debugging of data quality issues, when there's data corruption.

Continuous quality checks along the pipeline provide maximum reliability. Continuous checks, here, mean data quality checks as data flows through the pipeline. Since there could be multiple versions of the same record, continuous checks provide rigorous verification of quality for all or a majority of record versions, thereby minimizing issues that appear randomly, without any pattern. But if data volume is too high and continuous verification is impractical, quality checks can be done in a bulk manner. Data output within a specific time interval (with the latest version of a given record) can be selected in bulk and compared against bulk data input from the previous phase, for correctness.

For example, data output from extraction phase for records between 1pm and 3pm could be compared to data input from source data store, for the same time interval. An attribute/field that specifies this timestamp (like the last modified timestamp) can be used to extract required records. Data can also be sampled across the full data set, to reduce comparison volume. This sample could be based on a range of some column domain values or timestamp.

Challenges

The complexity of a data pipeline within any organization cannot be determined in a generic way, based on a fixed set of factors. Different parameters have different kinds of impact on the design and implementation of end to end data pipelines.

Design Aspects

Factors like transaction volume and data size typically have similar impacts in many data environments, in their contribution towards increased complexity. However, factors like data privacy, sensitivity that is different within different companies and vary with the type of data could make the design complex, even with manageable transaction volumes and data sizes.

For example, handling data with personal information like SSN or sensitive data like credit cards requires a lot more checks and security measures to be in place. This could contribute to more phases and access restrictions within the data pipeline from source to destination. In some cases, like CLOBs, Images, and NoSQL documents, data could have low to medium transaction volumes, but very large data sizes.

While overall data size volume may be manageable within extraction and other processes, downstream stores may have restrictions or performance degradation with large records. For example, Kafka's typically optimized for message sizes in low megabytes; performance typically degrades for large messages. Large messages slow down the brokers as they impose memory pressure on the broker. Similarly, in cases where the centralized repository is a custom hub using some database like MySQL, query performance could degrade for very large message sizes.

Operational Aspects

It's very important to thoroughly monitor and measure every part of the data pipeline. It's practically impossible to maintain and operate a data pipeline without clear visibility into its operational aspects. We'll delve into more details on operational aspects of data propagation and consolidation in a separate section; but below is the list of typical challenges that need to be addressed for any data pipeline to be practically useful.

1. Reliable Monitoring – Monitoring needs to be
    available for every component of a data pipeline.
    High-level monitoring of the entire pipeline, without
    granular health checks at the sub-component level,
    will not be sufficient to provide a robust and reliable
    data pipeline. Sub-component level monitoring
    is required to be able to debug and address issues
    effectively especially when the data pipeline is
    complex and/or involves multiple phases.

    Monitoring should be embedded within each
    component process, every program, job, SQL, etc.
    that needs to run to facilitate a functional data
    pipeline. Things can only be fixed when they're
    actively monitored and attended to. Delay in
    reaction time could complicate recovery process
    and recovery time from failures. Further, active
    monitoring on various aspects of functionality and
    supporting processes could allow for pro-active
    detection and prevention of failures. For example,
    detection of a continuously increasing load pattern
    on a data extraction host can help prevent complete
    system failure by pro-active action to mitigate or
    recovering from the issue.

2. Metrics – Metrics facilitate constant visibility into the
    functioning of the system. Some constituent process
    within the data pipeline could be running, without
    failure, without really accomplishing the intended
    goal. Without appropriate metrics, it's impossible to
    manage data flows efficiently and guarantee
    high availability of the pipeline to the consumers.

Loss of functionality detected too late can be hugely detrimental to consumers and the recovery process as well.

Metrics allow us to constantly track the functioning of various aspects and compare them to baseline metrics. Baseline metrics are the expected standard-state numbers for any system. For example, if an ETL system on top of a company's user database is expected to extract approximately 500 transactions/sec on Monday morning, this would be the baseline number for ETL metrics, for this database, on Monday. Any significant deviations can be tracked and acted upon. Sometimes, even if the deviation is not negative in terms of functional impact, it could result in further refinement of the baseline itself, probably due to variation in traffic, features, or some other factor.

Metrics also enable quantification of the impact of external factors on the core system. For example, percent change in transaction volume metric can quantify the impact of a new feature rolled out on the company's website or some other change in the application flow process. This kind of measurement enables clear visibility into impact and facilitates appropriate planning. It also provides the flexibility to ramp up changes and new features slowly and gradually based on impact across various system metrics.

Different environments could have different challenges but designing a data pipeline by considering key factors specific to the environment (data types, sizes, etc.) is very important to ensure the pipeline functions and scales well. Another important challenge is to make sure to provide operational visibility, through monitoring and metrics, with appropriate granularity and within individual components of the system.

Publishing to Kafka

As we discussed earlier, there are different ways to centralize and consolidate the source data extraction part of the pipeline. This promotes efficient reuse and distribution of source data across downstream data flows, and avoids redundancy in data collection and aggregation. There are few choices for the centralized data store model, from simple hub data store (flat files, database) to a highly distributed and scalable data store that provides scalability and support for multiple consumers, with parallelism.

Apache Kafka is a great option as a central data store for high-volume use cases and can provide streaming data for downstream consumption. Following details, extracted from Kafka's official web site, describe the key capabilities and features of Kafka. Apache Kafka is a distributed streaming platform. What exactly does that mean? What is Kafka good for?

We think of a streaming platform as having three key capabilities –

1. It lets you publish and subscribe to streams of records. In this respect, it is similar to a message queue or enterprise messaging system.

2. It lets you store streams of records in a fault-tolerant way.

3. It lets you process streams of records as they occur.

It gets used for two broad classes of application –

1. Build real-time streaming data pipelines that reliably get data between systems or applications

2. Build real-time streaming applications that transform or react to the streams of data

To understand how Kafka does these things, let's dive in and explore Kafka's capabilities from the bottom up.

1. Kafka runs as a cluster on one or more servers.

2. The Kafka cluster stores streams of records in categories called topics.

Each record consists of a key, a value, and a timestamp. Kafka has four core APIs –

1. The Producer API allows an application to publish a stream of records to one or more Kafka topics.

2. The Consumer API allows an application to subscribe to one or more topics and process the stream of records produced to them.

3. The Streams API allows an application to act as a stream processor, consuming an input stream from one or more topics and producing an output stream to one or more output topics, effectively transforming the input streams to output streams.

4. The Connector API allows building and running reusable producers or consumers that connect Kafka topics to existing applications or data systems. For example, a connector to a relational database might capture every change to a table.

In Kafka, the communication between the clients and the servers is done with a simple, high-performance, language agnostic TCP protocol. This protocol is versioned and maintains backwards compatibility with older version. Kafka clients are available in many languages. For more details on the anatomy of topics, partitioning, messages within Kafka, please refer to the official documentation or the immense wealth of information available on the Internet.

Let's focus on practical aspects of leveraging Kafka as a centralized data store, within the data pipeline.

In general, data can be published to Kafka in any format. Data can be pushed, through Kafka producer, in a delimited or even unstructured format to Kafka. But the structure/format of data pushed is very critical in facilitating reliable data consumption for downstream consumers. It's very important to minimize isolation between producers and consumers of data by publishing data in a standardized format. If the format is self-defined like Avro, it makes life a lot easier for data consumers since they can consume data along with its metadata, using a predefined schema. Data encoded with the schema is guaranteed to comply with defined data types and constraints defined within the schema definition.

Schema and Data

As we discussed earlier, one of the key advantages of using a defined schema is to enable enforcement of schema and constraint validation on data. Let's take the example of using Avro schema to publish data to Kafka.

Data extracted from sources is processed and encoded to Avro format. Avro serialized records are published to Kafka as events that can be consumed by downstream consumers. The granularity of events published depends on the specific use case. What this means is: same source data can be written at the same record granularity (or) after some level of aggregation, depending on the requirement. Same data with different aggregation levels can be written to different Kafka topics as well.

Data is written to Kafka topics that consumers can subscribe to. Kafka also allows for ACLs and thereby access security to be implemented on data topics.

Sample Schema

Data should be written to Kafka using a generic schema, for better scalability. If the requirement is to publish data from source MySQL databases to Kafka, as part of the data pipeline then let's assume the required granularity is at source record level.

If the source MySQL database has 1,000 tables, the Avro schema that's used to encode data before publishing to Kafka will be different for these 1,000 tables because the table structures will be different. Obviously, this could get complicated, with a greater number of data sources/tables.

An efficient way to address this problem is to use a generic schema, as shown below:

```
{
  "type" : "record",
  "name" : "generic_payload",
  "namespace" : "test.Kafka_producer",
  "fields" : [ {
    "name" : "table_name",
    "type" : "string"
  }, {
    "name" : "schema_id",
    "type" : "long"
  }, {
    "name" : "payload",
    "type" : "bytes"
  } ]
```

The schema above uses a generic format to publish data for all tables. The payload field takes the actual payload that will be different for different tables. For the payload schema specified by the "schema_id" field will be used. This schema will be different for different tables. Downstream consumers can reference the schema ID value and use it to de-serialize payload data.

Of course, this process flow also needs a place for consumers to get the actual schema corresponding to the value of the schema_id field. There needs to be a central repository for consumers to get schemas for different schema IDs. We'll get into more details on this schema repository, in a later section.

In general, the same generic payload can be used to publish payload1 with schema s1 for table t1 and payload2 with schema s2 for table t2. This also allows for schema evolution for a given table where schema s1 for table t1 could change to s1_new. Since the ID itself is included within the payload, downstream consumers can use the new id to get the new schema.

Schema Repository

As we discussed earlier, there needs to be a central schema repository for the generic payload and consumption process flow to work seamlessly. The repository could be a simple table and/or API that can be accessed by all consumers. The Avro schema corresponding to a data set will be published with an ID (could be simple hash of the schema itself) to the schema repository. This is the ID that will be embedded with the data (schema_id field value, from the sample schema provided).

As part of publishing data into Kafka, it's the responsibility of Kafka producers to keep this schema repository up-to-date with latest schema information. New schemas from new data sources and evolution of existing sources need to be persisted to schema repository before data is published into Kafka so that consumers can continue to consume data seamlessly. Of course, if the new schema is backward compatible, consumers can continue to access data even if the new schema is not readily available. In general, it's a good approach to make new versions of Avro schema for a given data set backward compatible.

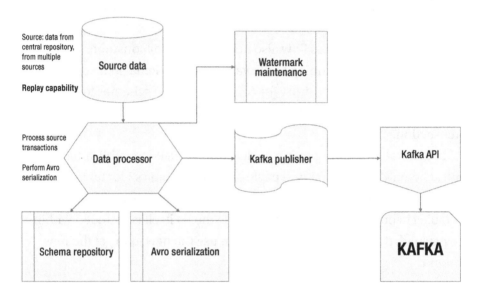

Figure 3-4. *Process flow from source data to data processor via Avro serialization*

Multiple Topics and Partitioning

Data published to Kafka is consumed from Kafka topics that provide parallel consumption capabilities to multiple consumers. Technically, this means you could publish all the data to one topic and ask all consumers to read from that one topic.

But there are practical limits and considerations to this approach. For example, data from 100 tables is published to the single Kafka topic in question. A specific consumer needs from just one of the 100 tables. This consumer may need to process a lot of data, corresponding to 100 tables, to get to its data of interest. This could introduce unnecessary latency for the consumer.

This pipeline could be made a lot more efficient by writing that one table to a different topic. This will enable faster processing and consumption for the consumer described above. In general, this approach of organizing required data within different topics, will help a lot in minimizing consumer latencies. This may not translate to a single table writing to a different topic, all the time; but the publishing strategy needs to be optimized to minimize consumption latency, without any consumer having to process and skip a lot of data.

Similarly, when it comes to parallel consumption, partitions within Kafka topics can be leveraged to further advance this optimization strategy. Data written to a topic can fe partitioned by specific field. For example, data within a source table has multi-tenant data for different clients; each client is identified by a client ID. In this scenario, the Kafka topic that receives data for this table could be partitioned by client ID so that different clients can be processed efficiently in parallel. Similar optimization can be achieved by smartly partitioning data across appropriate fields within source data sets.

Sizing and Scaling

In general, Kafka scales very well for heavy workloads because of its inherent distributed nature. But performance may start to degrade when the average size of data event goes up. Specifically, if there are events with large message sizes (> 1MB), they could adversely impact consumption across the entire topic.

There are a few settings that could be tweaked to allow for large message sizes; but like many other parameters, these comes with trade-offs as well. Worst case, one very large message could end up degrading the performance of an entire Kafka topic.

Practical workarounds need to be adopted for use cases with large message sizes. One of the solutions to this event size problem is to partition the large message into multiple small messages and re-assemble it during consumption. You could develop your own custom Kafka client (consumer and producer combination) that handles partitioning and re-integration that meets your requirements. Few open-source Kafka clients provide this partitioning capability.

Tools

The most convenient way to publish data to Kafka is to implement and maintain your own publishing code that leverages Kafka clients for the Kafka version used. There are appropriate versions of Kafka clients available for every version of Kafka. Companies that actively invest in Kafka infrastructure and contribute back to the open source community release their custom Kafka versions and clients. Though this approach requires the owner to implement and maintain code and infrastructure to manage publishing data to Kafka, it gives the most flexibility in terms of schema and data management, and possibly performance.

There are a lot of available tools that plug into source data stores and publish data to Kafka. They provide the ability to specify the Kafka setup parameters and have data published directly to Kafka topics.

However, they come with limitations on schema management, topic and partition management. The data pipeline will have to work around these limited capabilities to fit data flow into available options within these tools.

For example, Oracle GoldenGate provides a big data adapter plugin that publishes data from Oracle and MySQL tables directly to Kafka. This plugin provides the capability to publish to a single topic and one-topic-per-table. It also provides a generic schema with payload (similar to the one discussed above) and the ability to publish schema changes to a schema topic.

Tools like this provide instant ability to easily publish source data. But obvious limitations and trade-offs exist in terms of ability to spread data across topics and partitions in a flexible way. Further, the generic payload, currently, is rigid and lacks flexibility to add more descriptive fields like event identification timestamps, as required.

Similar tools exist for other data stores. But they almost always come with limitations in functionality and design. If what they provide is good enough, the best approach is to leverage them and simplify the data pipeline.

Conclusion

The chapter talks about the principles of change data capture and event streaming mechanism using Kafka. Throughout the chapter, we have maintained the fact that fundamentals and strategy are much more critical than execution. If we understand data sources well and how the changes have to be flown further to the consumers, we can plan an efficient and stable implementation.

In the next chapter, we are going to start off with the data processing strategies in a data lake. It will be of great interest for the readers who arrive from a database development background. Readers will understand the ways to work with the data in a data lake using Hive, Pig, and Spark.

CHAPTER 4

Data Processing Strategies in Data Lakes

In pioneer days they used oxen for heavy pulling, and when one ox couldn't budge a log, they didn't try to grow a larger ox. We shouldn't be trying for bigger computers, but for more systems of computers

—Grace Hopper, an American computer scientist

Data analytics trends have been disruptive. It would be an understatement to say that within the data analytics practitioner community, there exists a lean school of thoughts for data processing and drawing insights that are meaningful for business. With the steep increase in data appetite, data management practices have folded to multi times; which in turn has reinforced advanced analytics expertise and data management policies in the industry. The thought process behind crafting a data strategy is driven by use-cases and adjunct to technical capacity, learning momentum, and most importantly, the ability to cherry pick key discoveries that can be magnified into actionable insights to engage customers and drive

© Saurabh Gupta, Venkata Giri 2018
S. Gupta and V. Giri, *Practical Enterprise Data Lake Insights*,
https://doi.org/10.1007/978-1-4842-3522-5_4

business. The success mantra for a data analytics practice to excel is to maintain a "preamble" that envisions end goals aligned with the business use cases; both in the short run as well as the longer run. In our earlier chapters, we discussed the pillars of data analytics, i.e., data engineering, data discovery, data science, and data visualization. Data engineering offers relatively a bigger playground encapsulating ingestion principles, processing techniques, and development.

Approaches of data engineering encapsulate data ingestion as well as data processing. Ever since Google has published whitepapers on distributed frameworks, the gamut of data analytics has attained a definite maturity level through a series of sparkling innovations. These innovations have cut across different layers like data ingestion, processing, engineering approaches, and data representation. In the last two chapters, we tackled the challenges of bringing data into data lake. The focus of this chapter is to understand how to access data in a data lake. We will start with the classical parallel computing framework and cover key topics including datawarehousing in data lake using hive, data operations using pig, and modern-day processing framework i.e. spark.

MapReduce Processing Framework

In 2004, Google engineers Jeffrey Dean and Sanjay Ghemawat published a white paper[1] on distributed processing framework. The white paper illustrated a framework, operationally distributed, and aimed to process massive volumes of data. Here is what they quoted:

---

[1]Dean, Jeffrey; Ghemawat, Sanjay; MapReduce: Simplified Data Processing on Large Clusters, `https://static.googleusercontent.com/media/research.google.com/en//archive/MapReduce-osdi04.pdf`

"Our abstraction is inspired by the map and reduce primitives present in Lisp and many other functional languages. We realized that most of our computations involved applying a map operation to each logical "record" in our input in order to compute a set of intermediate key/value pairs, and then applying a reduce operation to all the values that shared the same key, in order to combine the derived data appropriately."

In simple words, MapReduce is a massively parallel processing framework for large data sets. Let us have a cursory look at the programming model. In a dedicated processing paradigm, data computation layer is static. During processing, data moves from the storage layer to the processing layer. On the other hand, in a distributed processing framework, the computation logic is pushed down to the storage layer. This means that a processing unit gets instantiated and each "processing" instance comes down to the node where data resides. The approach yields huge benefits from stability to the optimization perspective. An instance of program unit runs locally on the data node within a java container (JVM). It consumes a subset of data that resides locally on disk. Since the processing runs on multiple nodes in parallel, load gets spread out across the cluster; thereby greatly improving the performance. In addition, unlike in the past, the network never has a bottleneck because there is no data movement.

Motivation: Why MapReduce?

"The beauty of MapReduce is that any programmer can understand it, and its power comes from being able to harness thousands of computers behind that simple interface"

—David Patterson, a distinguished computer scientist.

Why do we need a distributed processing framework? The motivation is as sharp and simple as the paradigm. In a standard process-oriented architecture, two systems interact through messages (message processing interface). In a distributed environment, every node interacts via SEND and RECEIVE messages. Once issued, another node must receive and send back the acknowledgement bit. Not only does network-tied communication add to the latency but also, any loss in communication between the cluster nodes impacts cluster health. When working with huge volumes of data, network does become a bottleneck. Consider scenarios such as when an algorithm processes huge amounts of data or you need to sort petabytes and terabytes of data.

To prevent network congestion, processing can be pushed to the nodes of commodity cluster. With the subset of data on a node, a processing operation generates intermediate output files. A reducer operation works on these intermediate output files to aggregate or consolidate the final output. The "push down" operation is a step to assign a process instance to a data node and *map* values against the keys, while aggregation of values is a *reducer* operation on multiple intermediate outputs to final output. The operation comprising of *map* and *reduce* tasks is, thus, known as MapReduce.

The MapReduce flavor available until Hadoop 0.20 is considered the first version of MapReduce. In Hadoop 0.23, MapReduce framework had a resource management component known as YARN (Yet Another Resource Negotiator). We will cover YARN after a concept refresher on MapReduce version 1.

MapReduce V1 Refresher and Design Considerations

Figure 4-1 shows the components involved in a classical MapReduce operation.

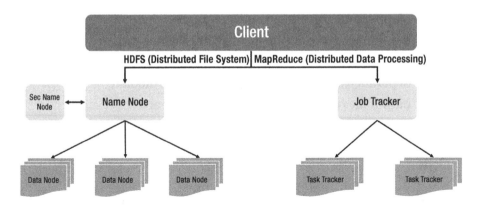

Figure 4-1. *A classical MapReduce operation*

As a quick refresher, let us focus on key highlights of the MapReduce framework –

1. Two broader phases – map and reduce. Input and output data format for both the phases is a key-value pair.

    a. Key-value based data distribution is crucial from a data access and availability perspective

    b. Key and value are java objects.

        i. Values implement *Writable.*

            1. Writable makes quick and easy serialization.

            2. IntWritable, LongWritable, FloatWritable, DoubleWritable, Text

        ii. Keys implement *WritableComparable.*

            1. WritableComparable can be compared to determine order.

2. Mapper jobs run concurrently on nodes where required data blocks are located. This avoids data transfer over the network and brings up the power of parallelism. Mapper jobs generate intermediate output files either locally on the nodes or at a defined location.

    a. Map is a stateless operation.

    b. Extend *Mapper base class*. LongWritable and Text are input key and value types. `Text` and `IntWritable` are output key-value types. Keys are `WritableComparable` while values are writable.

        ```
        public class SampleMap extends Mapper
        <LongWritable, Text, Text, IntWritable>
        ```

    c. Map method can be created for actual implementation.

3. Mapper class –

    a. Transforms input key-value pair to an intermediate pair based on the mapper class logic.

    b. Input and output file formats may differ in types and counts.

    c. One map task per input split generated by `inputFormat`. `InputFormat` is specified in driver code which contains –

        i.   Input data location over HDFS

        ii. How input split has to be created

iii. Creates RecordReader that parses data into key-value pairs before consumed by mapper

iv. TextInputFormat creates LineRecordReader object

v. Other standard input formats –

  1. FileInputFormat – Abstract base class for all file based InputFormats

  2. KeyValueTextInputFormat – Lines terminated with '\n' are treated as [key value] pair (tab separator)

  3. SequenceFileInputFormat – [key value] pair binary file

  4. SequenceFileAsTextInputFormat – [key value] as maps key.toString () and value.toString ()

4. Only after all mapper jobs are finished, reducer jobs kick off in parallel on the nodes.

  a. Mapper output can be stored in HDFS by writing output key-value pair using context method.

5. Reducer operations internally employ intermediate stages like combiner, and sort and shuffle to optimize the function.

  a. Single reducer or multiple reducers

    b. Extend *Reducer class* and override the *Reduce* function. Note the input and output key value pair parameters. Keys are `writableComparable` and values are `Writable`.

```
public static class Reduce extends Reducer
      <Text, IntWritable, Text, IntWritable>
```

6. Reducer class –

    a. Shuffle and sort phase starts after mapper phase is over and while fetching the input for reduce operation.

        i. Shuffle – collates all values associated with a key on a single machine

        ii. Sort – applies merge sort on input keys

    b. Reducer invokes aggregate function(s) (sequential) on all values associated with each key. Reducer output associated with a key, written to `RecordWriter` via a context object.

    c. Parallelism realized by concurrent reducer operations on multiple keys.

7. Map and Reduce classes must be packaged into a job to submit to the JobTracker.

8. JobTracker and TaskTracker

    a. JobTracker is a master daemon, responsible to assign mappers and track task execution progress

    b. TaskTracker is a slave daemon, runs on data nodes, and responsible to fire a JVM to execute mapper/reducer operation

9. MapReduce process architecture (Figure 4-2)

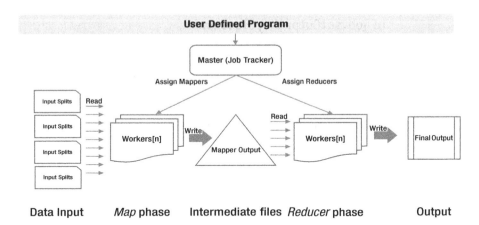

Figure 4-2. *MapReduce process oriented architecture*

10. SETUP and CLEANUP methods

    a. SETUP method is used to perform pre-mapper
        operations like initializing data structures,
        external data read, set custom parameters, or
        stats check. For the first time, it runs before map
        method is executed.

    b. CLEANUP is used to perform specific action
        after job operations get over.

11. Map-only operations

    a. Data sampling, ETL, or image processing
        require only mapper phase

    b. Set reducer count to 0

    c. `Job.setNumReduceTasks (10)`

    d. Use `job.setOutputKeyClass` and
`job.setOutputValueClass` in place
of `job.setOutputKeyClass` and
`job.setMapOutputValueClass`

    e. Therefore, `context.write` will write output Key-
Value pair on to HDFS and not on data nodes

12. Number of Reducers

    a. Single reducer by default

    b. If all keys in the final output, are required to be
sorted, use single reducer only.

    c. Single reducer may become a bottleneck if –

      i. Mapper output has multiple keys

      ii. Data volume ready for processing is hue

      iii. Good practice to partition keys of mapper

13. Compression – yield better performance by
compressing output after map and reduce phase.

    a. Set `mapred.output.compress` to TRUE to enable
compression at job output level

    b. Set `mapred.comperss.map.output` to TRUE to
enable compression of mapper output.

    c. Compression codecs supported – Java zlib, LZO,
bzip2.

14. Speculative execution – if a task is running slow,
Hadoop will try to run its multiple instances.
Whichever tasks finish fast, is considered a success
and the rest of all instances are suspended.

To enable speculative execution at map and reduce phase, set below two parameters –

   a.   `Mapred.map.speculative.execution = true`

   b.   `Mapred.reduce.speculative.execution = true`

15.   Limitations –

   a.   Restricted scalability – JobTracker runs on a single data node performing and responsible for tasks like resource management, task scheduling, and monitoring.

   b.   JobTracker exposes single point of failure. All mapper operations abort, if job tracker fails.

   c.   Suboptimal resource management –Number of mapper and reducer processes are fixed in advance for a tasktracker and cannot be leveraged flexibly. TaskTrackers cannot add mappers beyond a ceiling, even though unused reducer slots are available.

16.   Other frameworks that work along the lines of MapReduce

   a.   Microsoft Dryad – Directed graph based processing with programs as vertices and channels as edges. Multiple programs are connected via one-way channels.

   b.   Yahoo! S4 – it is a general purpose distributed streaming computing platform used for processing unbounded data streams.

   c.   Google Pregel

   d.   Twitter Storm

Yet Another Resource Negotiator – YARN

MapReduce 2.0 (MR2) or YARN was introduced in Hadoop-0.23 to revamp MapReduce V1 (MR1) by resolving its complications. In the classical MapReduce framework, critical factors like cluster utilization, resource management, and job monitoring were reallocated to find the best level. Unlike the MR1, a cluster must instead be treated as a resource grid wherein the resources can be allocated or released as and when required.

In MR2, resource allocation and job operation gets organized by maintaining a central body for resources management and job schedule. Figure 4-3 lists the benefits of using YARN.

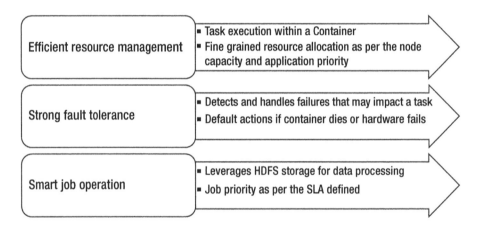

Figure 4-3. *YARN benefits*

For a cogent resource management, YARN architecture employs two components – global resource manager (RM) and application master (AM). The global resource manager is a central body that is primarily responsible to control resource utilization in the Hadoop cluster. The application master takes care of task operations for an application that are being executed on a data node. It is launched by global resource manager (RM) to supervise resources for a task and talk to node manager to schedule, execute, and monitor a task.

YARN ensures availability of task execution by checkpointing the status of applicationMaster. If need be, ApplicationMaster can reboot from the status from its last execution. Zookeeper helps in implementing failover mechanism from primary ResourceManager to backup RM.

YARN concepts

The way YARN works is that every MapReduce operation instantiates into a YARN application. Let's have a quick refresher of key YARN concepts –

1. Global ResourceManager – It is responsible for end to end resource management at the cluster level.

    a. Dumps its current state attributes in zookeeper for high availability

2. Two main components –

    a. Scheduler – does the resource allocation for applications. It receives and collates all requests from ApplicationMasters and further allocates the resources as per the priority and capacity.

    b. ApplicationsManager – manages job operations through ApplicationMasters. It ensures jobs are alive, optimized, and AM is through a smooth lifecycle.

3. NodeManager – Proxy agent that runs on each slave node and is responsible for launching application container, scheduling task in coordination with *Scheduler* and resource monitoring. The joint act of nodeManager and ResourceManager classifies into a stable data processing framework.

4. Application master – Cooperates with NodeManager
    on the direction of *ApplicationManager* to assign
    tasks to containers. It can be custom developed
    per application to fit the needs and coordinates job
    operations as per the defined SLAs.

5. Container – A container is a JVM that acts as a home to
    a task execution for an application. In MapReduce-1,
    the JobTracker used to assign slots for mappers and
    reducers; and these slots being fixed, didn't ensure
    optional resource utilization. A container replaces
    the fixed-slot approach by getting exchanged during
    task execution and enabling allocation based on
    parameters like memory and compute capacity.
    A *ContainerLaunchContext* indicates an object
    containing resources required to launch a container
    along with the commands to be executed.

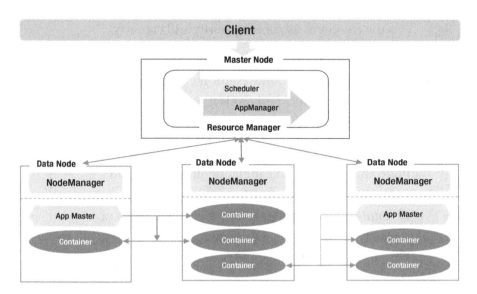

Figure 4-4. *Subcomponent level design of YARN*

YARN application execution goes through the below steps –

   i.   Client submits an application to Resource Manager (RM).

   ii.   ApplicationManager daemon starts and registers with RM. It estimates the resources required by the application.

   iii.   RM coordinates with NodeManager (NM) to launch ApplicationMaster (AM) on one of the nodes.

   iv.   AM gathers request details from NameNode and submits resource request to RM for task execution.

   v.   RM queues the request from AM until the resources get freed up on worker nodes

   vi.   AM receives containers from RM to execute application tasks on specific hosts (slave nodes).

   vii.   AM coordinates with NodeManager to assign task to the containers. NodeManager helps with the task scheduling and resource monitoring.

   viii.   After successful execution, AM confirms the job status to global resource manager.

6.   Resource model and resource negotiation highlights:

   a.   Multiple containers sized 512M or 1G can be started on each node.

b.   Per application ApplicationMaster requests for containers depending upon resource requirements, subject to the capacity limits for an application.

c.   In its protocol, ApplicationMaster specifies hostname, resource requirement, container, and priority.

d.   In MR2, cluster resources are not split into mapper or reducer slots.

e.   RM's scheduler keeps an eye on cluster resource utilization and prevents resources from getting over allocated by checking the limit metric for an application, user, or queue.

f.   Resource Monitoring – NodeManagers send out the resource usage metric to RM Scheduler.

7.   YARN implementation highlights are:

a.   Start a YARN client of YarnClient type

i.   YarnClient.createYarnClient();

b.   Create application of `YarnClientApplication` type

i.   testYarnClient.createApplication ()

c.   Set application context

i.   testYarnClientApp. getApplicationSubmissionContext;

d.   Set resource requirements for an application context using Resource object

i.   Resource.newInstance (memory, cores);

    e.   Launch a container of ContainerLaunchContext object

        i.   ContainerLaunchContext.newInstance (localResource, environment, byteBuffer Tokens, application acls)

        ii.   applicationContext.setAMContainer ([ContainerLaunchContext])

    f.   Set application priority of Priority object

        i.   Priority.newInstance (priority)

        ii.   applicationContext.setPriority ([Priority])

    g.   Submit application

        i.   testYarnClient.submitApplication (testAppContext)

8.   YARN web UI provides a more detailed view of task execution than Hue. It runs on ResourceManager host which becomes its entry point as well on UI. Note that none of the YARN execution can be controlled from UI.

9.   YARN doesn't keep track of job history. Spark and MapReduce provide job history server that archives job details (metadata and metric) and can be accessed via Hue or its own UI.

Hive

The fact is widely accepted that Hadoop platform has well addressed the challenge of data volume; be it storage capacity or data processing. The manifestation of data has changed drastically since the emergence of non-traditional data sources. Data centric challenges are not just volume and velocity, but processing and operational processing add to the complexity of so called "data challenge".

Most of the data processing layers are written in low level language. There were growing schools of thought which highlighted the fact that primary implementers and executors of data processing layer are data development and analyst professionals who may or may not be handy with java or other low-level languages. Database professionals feel at home with a structured data processing language. This need led to the conceptualization of query languages that can retrofit the Hadoop processing layer and provide a layer of abstraction over MapReduce.

The project Hive was started by Facebook in 2007. It provides a solution to non-java practitioners who perform data warehousing over hadoop. Hive employs a SQL like declarative query language, HiveQL to communicate with Hadoop cluster. Its close resemblance with SQL semantics remains heavily responsible for its wide adoption amongst the data analyst community.

Hive was created for scalability and ease of use for data professionals. The performance of a Hive query might be much slower than expected as low latency never fell under the list of objectives for Hive. It is not designed for online transaction processing. However, with the most recent works like Tez and Live Long and Process (LLAP), Hive performance can be optimized for low latency requirements. We shall discuss these topics later in the chapter.

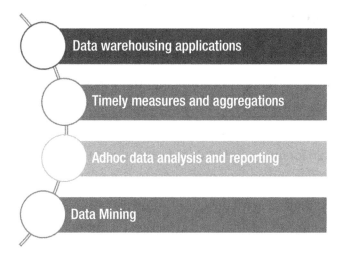

Figure 4-5. *Data-centric capabilities of Hive align with key data warehousing objectives*

Hive – Quick Refresher

In this section, we will have a quick refresher of Hive architecture, abilities, and concepts. Since Hive forms a layer of abstraction on top of MapReduce, all HiveQL statements are converted into MapReduce jobs which are then executed on the Hadoop cluster.

The anatomy of a standard query states that it must have data sources in a structured format and not in flat files. Hive has the ability to give structure to various data formats, i.e., *Schema on Read* to unorganized data sets. This is achieved through Hive metastore that acts as a catalog containing table definitions.

Hive Components

Below is the list of components involved in a query execution through hive.

1. Interactive shell – environment from where the query is initiated.

2. Hive Driver – receives the query requests and is liable for query operations like creating a session handler and passing over query to compiler.

3. Compiler – Parses the query, looks up to metastore for object validation, and generates an optimized execution plan.

   a. Consists of parser, semantic analyzer, logical plan generator, and query plan generator

      i. Semantic analyzer does sanity check for query by accessing metastore for table definition and column properties.

      ii. Plan generators responsible for converting execution plan to MapReduce tasks

   b. The optimization includes performance analysis of different query blocks and achieve plan transformations. For example, metastore lookup can reveal tables' partitioning strategy and query predicates can determine if partitions can be pruned out or not. Similarly, implicit data type conversions are also taken care of at compilation level.

   c. Execution plans a direct acyclic graph (DAG) of operational stages involved in query execution.

4. Execution engine – physical operations to fetch data and prepare query result set.

5. Hive metastore – It is the nucleus of Hive operations. It constructs a layer of data abstraction and the ability to distill down data variability into defined shapes and structures i.e. schema and table definitions.

   a. HiveServer or HiveServer2 provides thrift interface for external applications to connect via JDBC/ODBC.

The architecture diagram in Figure 4-6 shows the connection between different components of Hive.

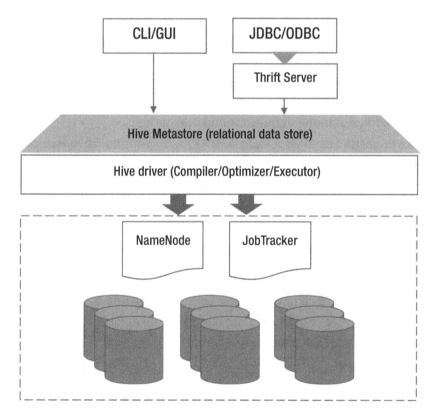

Figure 4-6. *Hive architecture*

Hive Metastore (a.k.a. HCatalog)

Hive metastore is the central schema repository of Hive query system that stores metadata of Hive data models (tables, partitions, and buckets), serializers and deserializers, and information about HDFS file location (Figure 4-7). A relational database store serves as the backend of a Hive metastore. The backend piece is implemented by an object-relational mapping solution called Data Nucleus. Hive metastore can be packaged with relational databases like Derby, MySQL, SQL server, Oracle, and Postgres. Relational data paradigm catalyzes the reliability and brings the ability to query metadata.

Metastore is tightly coupled with the Hive service or a high query processing system. As soon as a table is created in Hive shell, its definition gets stored in Hive metastore immediately. This way Hive service ensures the sync between data and metastore. On the other hand, Hive metastore can be leveraged by other processing frameworks like Pig and Spark and reuse predefined schema definitions and data models.

Hive metastore exposes a *metastore API service* for the Hive service and external clients. Hive service uses metastore service to store table and partition metadata in the metastore. External clients invoke metastore service to access metastore for table or partition information during a query execution.

A Hive metastore can operate in the three modes below –

1. Embedded metastore – All three services – hive, metastore API, and database service run inside a single JVM container. It uses derby as the default database. Only one Hive session can be opened at a time.

2. Remote metastore – All three services – hive,
    metastore API, and database service run in
    separate containers. External clients can connect
    to metastore via *thrift* service. To setup a remote
    metastore, metastore URIs must be configured at
    Hive service level (set `hive.metastore.uris =`
    `thrift://host:port`).

3. Local metastore – Hive and metastore API service
    run within a single JVM. Its underlying database
    is MySQL which runs in a separate JVM. Multiple
    metastore API requests can connect to database
    using JDBC or ODBC driver libraries. Please note –

    a. Driver libraries are available in Hive's classpath.

    b. `javax.jdo.option.ConnectionDriverName` to
        `com.mysql.jdbc.Driver`

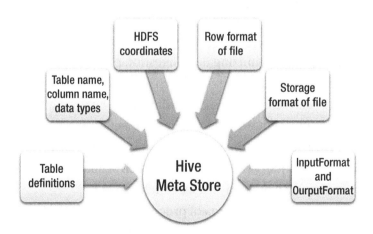

Figure 4-7. *Hive metastore serves more than the table metadata*

Hive – Design Considerations

Abiding by its pro-data warehouse principles and mimicking SQL semantics, Hive finds wide application in ad-hoc reporting, log processing, text mining, predictive modeling, business intelligence, and analytics. In an enterprise data lake, Hive occupies a critical spot as data analysts and data scientists, who eventually emerge as primary consumers of data, are extremely fond of SQL to speak data. MapReduce stands well from a pre-determined application standpoint but ad-hoc data centric exercises in a java can be nightmarish. Before we jump on to the bulleted factors that impact Hive design and development, let us understand the relevance of exercises like partitioning, bucketing, and denormalization in hive.

1. Partitioning – Partition works on divide-and-rule mechanism by reducing the amount of I/O by a significant margin during data processing. In a traditional data warehousing environment, processing performance becomes an equation of data modeling and tuning practices like indexing, statistics, etc. Hadoop has no place for indexes and that helps in optimizing data ingestion pipelines. However, it also implies that each and every query has to scan and read the entire data set, even though the requirement was really a tiny data set. With growing data sets, the table scans may negatively impact the query performance. In such scenarios, partitioning could play a sheet anchor role by reducing the surface area of processing and allowing space for other data-centric operations. In a Hadoop cluster, partition or a sub-partition will be present as directories within the table directory. If the query contains predicates using partitioning key column,

chances of partition pruning grow even brighter. Keep this in mind while electing the partitioning key to avoid small file handling in hadoop.

2. Bucketing – Like partitioning, bucketing is another way of handling a large data set by slicing it down in evenly sized subsets. If you have heard and worked with hash partitioning, bucketing is a very similar concept. Bucketing not only helps in even distribution, but also while joining more than one table. If tables participate in a join based on bucketing key, the bucketed connect in a much simpler and optimal fashion. If the size of buckets being joined is small, chances of a map-side join in the mapper phase increases. Reducer joins, being resource intensive, should generally be avoided.

3. Once again, keep an eye on the resultant files and ensure that bucketed files do not become too small as to allow slippage of cluster resources. An optimum bucket size should be a multiple of HDFS block size or as an exponential of two.

4. Denormalization – This is a technique to produce flattened data structures that hold maximum information required for data processing. In hadoop, query joins are the most resource intensive operations, within which reducer join is the most notorious one. If a data set can be modeled, de-normalized, and refreshed periodically in curated data layer, it can potentially avoid expensive joins. While partitioning and bucketing help in slicing down large data sets to support parallel operations,

denormalization helps in avoiding expensive query joins in data lake. Denormalization can be as simple as projecting columns from multiple tables jointly, or have aggregation and derived column at the table level.

Below are the key highlights of practical Hive from the design and development space.

1. Hive interface options are:

   a. Command line

      1. Hive shell

      2. Beeline shell

   b. Web UI

      1. Ambar

      2. Hue

2. Hive data models – Hive data model consists of databases, tables, partitions, and buckets

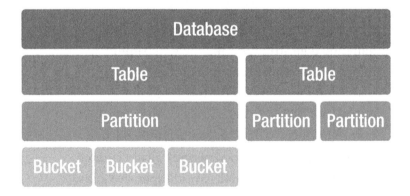

Figure 4-8. *Hive data model*

3. File handling best practices – supports parquet, text, Avro, RCfile, and sequenceFile file formats.

   a. For landing, Avro is preferred

      1. JSON can be used but requires dedicated SerDe processing

   b. For staging, use parquet for columnar data processing

      1. Use sequenceFile or RCfile for row oriented data processing

   c. For publishing, parquet is used with Hive or impala for better performance

4. Use native data types for columns that can be mapped to a native type in java. Large object data types (BLOB and CLOB) are not supported.

   a. Complex data types like maps, arrays, and structs are supported

5. Hive supports views, but no support for subqueries.

6. Hive supports aggregate as well as window analytical functions.

7. Hive support for DML

   a. Supports insert, update, and delete

   b. Below parameters need to be set in Hive configuration file and restart the server

      1. `set hive.support.concurrency=true;`

      2. `set hive.enforce.bucketing=true;`

      3. `set hive.exec.dynamic.partition.mode=nonstrict;`

4. `set hive.txn.manager=org.apache.hadoop.hive.ql.lockmgr.DbTxnManager;`

5. `set hive.compactor.initiator.on=true;`

6. `set hive.compactor.worker.threads=2;`

c. Only transactional tables are updateable. You must create table with an additional storage clause - `TBLPROPERTIES('transactional'='true');`

8. SORT BY, ORDER BY, DISTRIBUTED BY, CLUSTER BY

a. If more than one reducers are available, SORT will sort data in *each* reducer. This will not guarantee if the final result set will be sorted or not

b. ORDER by employs a single reducer to ensure the final result set is ordered.

c. DISTRIBUTE BY distributes the rows amongst reducers based on a key column. However, this is nowhere related to clustering or sorting actions. It is useful while distributing data amongst the reducers. All rows with the similar pattern will be assigned to a single reducer.

1. DISTRIBUTED along with SORT will sort rows within each reducer *(cluster by)*.

2. DISTRIBUTED along with ORDER will sort the final output.

    d. DISTRIBUTED along with SORT BY constitutes a CLUSTER BY operation on a single key column. All rows with the same cluster key will be assigned to a single reducer. In case the distribution and sort keys are different, you need either explicit grouping or DISTRIBUTED + ORDER clause.

9. In order to batch multiple rows together for processing, set the configuration parameter `hive.vectorized.execution.enabled` to true.

10. Hive partitioning

    a. Partitioning enables splitting of large volumes of data into small chunks depending upon partitioning key column.

        1. Support for horizontal partitioning for both hive-managed and external tables

        2. Partitions stored as sub-directories within table folder.

    b. Partitioning key should be a column with a controllable cardinality. If the cardinality of the column is very high (like timestamp), it will create too many directories within the system and may lead to data fragmentation issues.

    c. Partitioning helps in data organization and refining the data traversal at the input path level. Queries that use partitioning key in predicates are helped by *partition pruning* that greatly optimizes the query performance.

        1. Queries that do not make use of partitioning key go for a full scan.

d. Partition key column should not be specified in the table column specification. It gets displayed though, while describing the objects.

```
CREATE TABLE dataLake_sor
   (
     sorId    INT
    ,sorName STRING
   )
PARTITIONED BY (userId STRING)
ROW FORMAT DELIMITED
FIELDS TERMINATED BY '\t'
```

e. Static partitioning indicates that partitioning key column value was pre-known and used while loading.

f. Dynamic partitioning helps when moving data from non-partitioned table to a partitioned one. Instead of running load for each distinct value of partitioning key column, column name can be directly used.

g. Dynamic Partition Insert – if the partition key columns are already available in the source table, then explicit partition value can be skipped during inserts.

1. For every distinct value of the partition key, a partition will be added automatically within the RDS table.

h.   Below parameters control the *Dynamic Partition Insert* feature –

    1.   `hive.exec.max.dynamic.partitions.pernode` *(default 100)* - Maximum number of dynamic partitions that can be created by any given Mapper or Reducer

    2.   `hive.exec.max.dynamic.partitions` *(default 1000)* - Total number of dynamic partitions that can be created by one HiveQL statement

    3.   `hive.exec.max.created.files` *(default 100000)* - Maximum total files created by Mappers and Reducers

11.   Hive bucketing or clustering

a.   It is an optimization technique to distribute table data evenly amongst multiple buckets and control wide data scans. Bucketing operation is carried out by series of MapReduce jobs.

b.   Buckets must be created based on a column that has relatively high cardinality and can assure evenly data distribution.

c.   Buckets are created and stored as subdirectories within the table folder. Data is assigned to a bucket based on the hash value of the bucket key column.

d.   Bucketing helps in logical data organization and optimizes the queries that use clustering key column in predicates or joins.

e. Set below two parameters before inserting data into bucketed table

1. SET `mapred.reduce.tasks` = <no. of buckets>

2. Set `hive.enforce.bucketing=true`

f. SET `hive.optimize.bucketmapjoin=true` will direct Hive to leverage a bucket level join during map stage join

12. Hive partition and bucketing used together

a. Buckets can be created within partitions to optimize data storage and hence, the data scans within partitions.

b. For example, the below table will create partitions by YEAR and within each partition, there will be 25 buckets of areas containing their population details.

```
CREATE TABLE city_population_store
    (
    area STRING,
    record_date STRING,
    last_count INT,
    current_count INTO
    )
PARTITIONED BY (year STRING)
CLUSTERED BY (area) INTO 200 BUCKETS;
```

c. Once again, partitioning must be done on a column with low cardinality.

    d. If the clustering key column has low cardinality (less than the number of buckets specified), there are chances that buckets will remain unused (i.e., no data).

13. TABLESAMPLE can be used to pick up portion of data as a sample for query processing.

    a. Block sampling - Sample can be specified as a percentage of total rows or a finite count of rows.

For example, in the below query the input size of 5% or more of abcTelecom_logs will be used in the query processing.

```
SELECT *
FROM abcTelecom_logs
TABLESAMPLE (5 percent)
```

Sample can also be specified as 1000 rows or 100M.

    b. Bucketized sampling – Buckets can be created on a designated column or randomly to prepare sample data for query. Keep note that this query can run even for a non-clustered table, it is not very efficient though.

```
SELECT *
FROM abcTelecom_logs
TABLESAMPLE (BUCKET 5 OUT OF 50 ON city)
```

If abcTelecom_logs would have been clustered on CITY column, bucketized sampling would have been more efficient.

Hive LLAP

In an enterprise data lake ecosystem, data visualization tools like Tableau, Qlikview, or SiSense connect to Hadoop data lake and users can slice and dice the data set to gain insights in real-time from various gradients. Query performance heavily impacts the real-time experience and decision-making ability. Hive, as we know, forms an abstraction layer on top of MapReduce processing layer. It cannot afford to provide the degree of performance that a relational database SQL can promise. One of the latest trends to beat performance is to marinate caching technologies with disk operations. This is what, in a nutshell, Hive LLAP is all about. LLAP stands for Live Long and Process.

LLAP joined Hive 2.0 to warrant for query performance in Hive with Tez without compromising the native features of Hive. Keep in mind, it doesn't replace the existing execution model of Hive; instead complements it to fit the bills of faster data processing. Structurally, it consists of a daemon and a DAG based framework, orchestrated by Tez execution engine.

LLAP daemons run on YARN in a standard Hadoop cluster. It consists of two components, namely, *query executor* and *in-memory cache*. Query executors control some of the essential stages of query execution like query processing, query fragment execution, pre-fetching, access control, metadata, and result caching. Fragments that are lined up executed by their priority are capped as *queue fragments*.

Tez and Hive client coordinate LLAP engagement by determining which query needs to be pushed to LLAP daemon for execution.

Hive facilitates the processing and *query fragments* like HDFS location and metadata. YARN, clearly, remains the caretaker of resources for LLAP daemons. Query fragments are nothing but the bits required for query processing like operators, metadata, expressions, primary data types, and input and output channels. Operators in reference are joins, SQL clauses, and scan methods. Expressions could be the functions of all nature (SQL, scalar, numeric).

So, what queries can run in LLAP? The decision of running queries in LLAP can be configured in Hive client as all, none, or hybrid. In auto mode, preliminary criteria are data source (HDFS), file format (ORC), and data size. Small queries can be processed directly by the daemon. However, for large queries, YARN takes over as usual and drives through normal query processing stages.

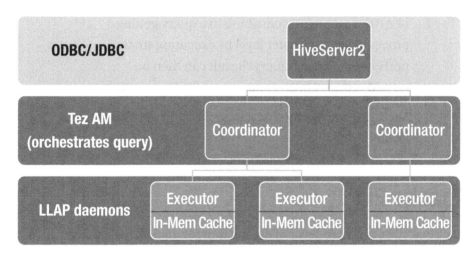

Figure 4-9. *Hive LLAP architecture*

Design considerations are:

1. Hive LLAP is available only via Tez execution engine.

2. Daemons are extensible through APIs to all such processing frameworks (Spark and Pig) who prefer relational views of data over file-based processing.

3. Currently, only ORC data qualifies to be cached. Rest all file formats can still be processed in LLAP daemons but cannot be cached.

4. Cache eviction policy – configurable eviction policy for different type of workloads.

5. By default, low latency queries are favored over heavy and complex queries. Queries keep waiting in a queue unless priority is specified.

6. LLAP daemon failure - Since the LLAP daemons are not primary executors, Tez AM can process query fragments on other data nodes of the cluster.

7. LLAP daemons can contribute to a query getting processed at the cluster level by executing small portions of a bigger query. Result can then be passed on to main Hive query or a further to a task.

8. Security – fine grained access security up to column level.

Apache Pig

Like Hive, pig provides a layer of abstraction for data processing on Hadoop. It provides an alternative to write MapReduce programs in low level language by facilitating a scripting layer that gets translated into instructions which follow execution on HDFS cluster. Pig simplifies the MapReduce layer by reducing thousands of lines of code to mere countable instructions.

Pig uses high level language Pig Latin to design data transformation and flow expression. It is the language interpreter which converts Pig Latin instructions into MapReduce or spark jobs and submits them for execution. The Pig engine consists of a parser, optimizer, and a distributed query execution. A program in Pig sustains an analogous data-centric approach as each and every instruction is connected to data streams.

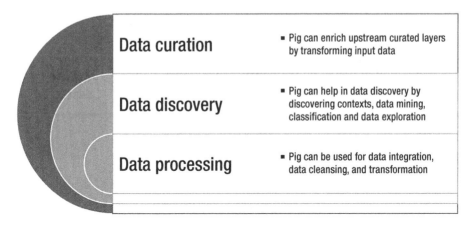

Figure 4-10. *Capabilities of Pig*

Pig eliminates the challenge of breaking down computation into map and reduce phase. If a problem can be expressed as chain of tasks or Direct Acyclic Graph (DAG) using standard operators and clauses like filters, joins and aggregation, Pig stands out due to inherent advantages of swift development cycle.

Though procedural, Pig uses SQL-like declarative constructs to develop scripts for data processing. It may not be as optimal as spark or Hive, but given the factors like flattened learning curve and fast development, it is preferred for its low-cost development, flexed out data model, and quick results.

Pig Execution Architecture

The following diagram represents the execution architecture of Pig processing (Figure 4-11).

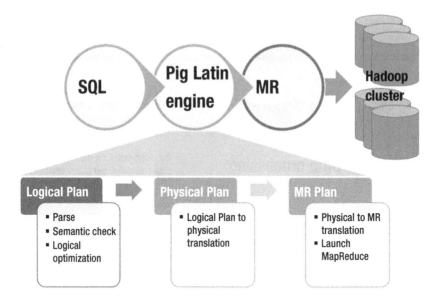

Figure 4-11. *Pig execution flow*

From the above architecture diagram,

1. Logical plan – It is an interim representation of task chain as sequenced in the script. Pig uses ANTLR parser (Another Tool for Language Recognition) to conduct semantic and sanity check of script, following which it generates an Abstract Syntax Tree (AST). Once AST is converted into logical plan, certain optimization opportunities are implemented like column pruning and process pushdown.

2. Physical plan – Logical plan is converted into multiple physical plans indicating bifurcation of physical computations for each logical operator. For example, JOIN is broken down into LOCAL REARRANGE and GLOBAL REARRANGE along with PACKAGE.

3. MR plan – Physical plan gets converted into
    MapReduce jobs with phase identification and data
    inputs and outputs. Physical computation steps
    are branched out in map, combiner, and reduce
    phase. In this stage, MR jobs are also inspected from
    optimization perspective. For example, mappers can
    be consolidated to reduce data transfer or modifying
    the number of reducers. Once frozen, MR jobs are
    submitted to Hadoop cluster for execution.

Pig provides EXPLAIN operator to generate the detailed execution plan
of a script.

Key concepts and design considerations

1. Pig data model – consists of atoms, tuples, bags,
    and maps.

    a. Atom is a scalar value of a primitive data
        type like INT, FLOAT, LONG, DOUBLE, or
        CHARARRAY.

    b. Tuple is analogous to a row from relational data
        world. It's a sequence of attributes enclosed
        within parenthesis.

    c. Bag represents a group of tuples and is
        analogous to a table from relational data world.

    d. Map represents a flexi-schema data structure of
        key-value pairs.

2. Input data stream model – no schema required
    while importing data into the script. As long as
    Pig execution engine can identify tuples in the
    data stream, there is no need to flatten down the
    HDFS files.

3.   Although Pig doesn't require schema at runtime, type conversion can be critical during data buffering and processing. Pig may predict the data type of an atom or tuple attribute based on the nature of processing logic.

4.   Extensibility – Pig supports user defined functions which are designed for specialized processing of data.

5.   Pig can be installed on an edge node that has connectivity to big data lake. Since Pig core is built on java, it becomes the primary pre-requisite to run Pig.

   a.   If Pig uses UDFs, the native compiler must be available on the host. For example, python or javascript based UDF must have the python and java script component installed.

   b.   For testing integration and build automation, respective components like Ant, Chef, or Junit must be installed.

6.   Pig run modes

   a.   Local – Connects to local file system where developers can play around with scripts, debug, and test features. No parallelism realized in local mode.

```
$ pig -x local
```

   b.   MapReduce – Connects to Hadoop cluster of data lake. Pig scripts can be deployed and run on wide data. Note that only query execution can be parallelized on cluster nodes. Query sanity and semantic operations are still carried out locally.

```
$ pig -x MapReduce
```

7. STORE vs DUMP – Pig follows lazy execution approach during script execution. Unless output action is not encountered in the script, the engine pipelines all statements in memory. Output actions can be either of a diagnostic operator – STORE, DUMP, ILLUSTRATE, EXPLAIN, or DESCRIBE, which indicate the creation of physical and logical execution plans to kick-start Pig Latin code execution. Bag scanning instructions may cause spill over to the disk.

    a. Use STORE, if the script result needs to be written on HDFS.

    b. Use DUMP, if the script output just needs to be displayed on command line.

    c. For debugging purposes and quick testing, developers can also use ILLUSTRATE which applies a sampling algorithm to create small test data, apply transformation, and achieve performance.

8. PARALLEL – Maneuver reduce phase of clauses such as ORDER, DISTINCT, JOIN, GROUP, COGROUP, and CROSS by specifying reducer count in PARALLEL clause. Reducer count can also be set at the script level by including the command below –

```
SET default_parallel [reducer_count]
```

Note that statement level parallel setting will override script level configuration. For map phase, parallelism becomes the function of input data splits. If reducers cannot be explicitly added to the script, you can modify the two two parameters below to tune reduce operation.

  a. `pig.exec.reducers.bytes.per.reducer` – Number of input bytes per reducer. It is set to 1G by default. For large data sets and depending on operation, this can be increased beyond 1G.

  b. `pig.exec.reducers.max` – Maximum number of reducers. By default, it is 999.

9. SAMPLE – The SAMPLE operator, allows script to play with subset of data, can be very useful in data profiling works. Scripts in Pig Latin can contain logic to perform data quality checks and build data profile.

10. Pig can be used to load data streams into hadoop. They are benefitted by the ability to scale by node count on Hadoop cluster.

Apache Spark

Spark is a scheduling, monitoring, and distributing engine that provides a next-gen processing framework just like MapReduce. It has slowly started replacing classical MapReduce models and has become a defacto in the big data world. Spark came out of the University of California, Berkeley's AMPlab project in Jan 2011. Reynold Xin, the Development Lead at Berkeley AmpLab Shark, quotes about Spark as – *"Spark … is what you might call a Swiss Army knife of Big Data analytics tools."*

Why Spark?

The MapReduce framework has been a trustworthy implementer of distributing computing for large volume data. It helped in solving complex problems through a series of mapper and reducer stages, distributed over the cluster, and executed in coordination with the distributed storage platform. What it mainly dealt in was the large on-disk data sets, which undoubtedly was great for batch processing, but not so well for low-latency models. One of the key traits of MapReduce processing is the ability to store interim as well as final datasets on cluster. MapReduce performs a lot of reading and writing (I/Os) to the disk throughout the transformation. This could be an expensive operation as it incurs both the replication of a dataset in the disk I/O and the network I/O as it starts a new JVM for each task which takes time for loading JARs and parsing XML configurations. Developers building MapReduce framework for a problem, need to code manually, which could turn into a cumbersome exercise, given the complexity of the job. Apache Spark overcomes these issues by introducing a completely different processing model through a combination of batch, streaming, and interactive computation.

Spark, an open source framework was initially started by Matei Zaharia at UC Berkeley's AMPLab in 2009, and open sourced in 2010. In 2013 the project was donated to the Apache Software Foundation. It can be used for processing humongous volumes of data in a data lake environment, hosted on premise or cloud. It offers developers an application framework that works around a centralized data structure. Keep in mind that spark is designed to enhance the computational speed, also covers wide range of workloads for example batch, interactive, iterative, and streaming.

Spark's approach towards processing has been largely influenced by Microsoft's Dryad [2]paper on parallel and distributed execution. It

---

[2]Dryad: Distributed Data-Parallel Programs from Sequential Building Blocks [https://www.microsoft.com/en-us/research/wp-content/uploads/2007/03/eurosys07.pdf]

introduced an in-memory caching abstraction that makes it ideal for workloads where multiple operations try to access the same input data. A user can cache data sets in memory which overcomes disk I/O overheads. Spark maintains an executor JVM on each node so launching a task regardless of MapReduce operations comes down to making a remote procedure call (RPC) to it and passing a runnable to a thread pool.

Before Spark 2.0 came into practice, the main programming interface of Spark was the Resilient Distributed Datasets (RDD). After Spark 2.0 it got replaced by Dataset which is strongly written like RDD, but with higher optimizations under the engine. Spark provides interactive shells in Python or Scala, it helps a simple way to learn various functionalities of Spark API.

Spark core and its member libraries form the building blocks of Spark stack. The libraries are optimized to fit into all stages of data management. For data integration, Spark streaming fits the bill. For data science requirements, Mlib and SparkR are available. For graphical processing, GraphSX is part of the stack. For data engineering, SparkSQL can be used by data analysts. Spark core acts as the brain of the stack. Figure 4-12 shows what constitutes the core engine of Spark.

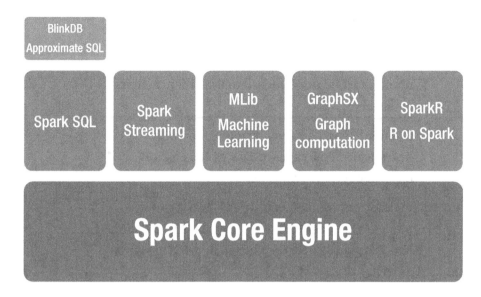

Figure 4-12. *Spark Stack*

Resilient Distributed Datasets (RDD)

RDDs are immutable collections distributed across the cluster that are created via data transformation and can be cached across parallel operations. The resilient distributed datasets are fault-tolerant as they are oriented in a graphical structure which makes it easier to re-compute partitions that got damaged during node failures. Let us check key characteristics of RDD –

1. Lazy evaluation – Data in a RDD is not populated until an action is triggered

2. Fault-tolerance – RDD preserves lineage information which helps in rebuilding lost partitions.

3. In-memory and cacheable – RDD resides as a memory structure restricted by the retention time and capacity of the memory. These datasets remain cached until flushed to the disk to achieve persistency.

4. Immutable or read-only – An RDD can only be transformed to a new RDD, which ensures consistency of the datasets.

5. Parallel, partitioned, and typed – RDDs can be processed in parallel as they are logically partitioned and distributed over the cluster. Records are typed as well.

6. Quite importantly, RDDs can be owned by one and only one SparkContext. They cannot be shared between more than one SparkContext.

7. An RDD can be made persistent by using `persist ()` or `cache ()` methods at different storage levels like disk or memory. A node can persist the partition of an in-memory dataset that it computes for future operations. This gives tremendous performance boost for the future actions on the dataset.

An RDD can be created in two ways –

1. Parallelize a collection - Parallelized collections are created by calling SparkContext's parallelize method on an existing collection in your driver program. Here in the parallelize method second argument denotes the number of RDD partitions in memory.

```
val data  = Array(1, 2, 3, 4, 5)
val distData = sc.parallelize(data, 4)
```

2. Spark can create distributed datasets from any storage source supported by Hadoop, including your local file system, HDFS, Cassandra, HBase, Amazon S3, etc. It supports files as per Hadoop inputformat like text files, SequenceFiles. For example, the below code snippet creates an RDD using SparkContext's textFile method.

```
val distFile = sc.textFile("data.txt")
distFile: org.apache.spark.rdd.RDD[String] = data.txt
MapPartitionsRDD[10] at textFile at <console>:26
```

An RDD can either go through an *action* or a *transformation*. Transformation creates a brand-new dataset from an existing RDD, while an action on a dataset runs a computation logic and returns a value. A map function on an RDD is a transformation as it passes a dataset element to return a new RDD, while reduce becomes an action that performs aggregation of RDD elements. Similarly, operations like filter, sample, union, join, cache, groupByKey, or reduceByKey transform an RDD into a new RDD. A reducer, collect, count, or save are the actions which will return a result to the driver. An RDD transformation is always a lazy evaluation which gets triggered on an action.

RDD Runtime Components

A Spark application can be run in a distributed mode. The resource management can be done either by spark cluster manager or YARN. However, it is always better to integrate it with YARN which has better knowledge of data locality on the Hadoop cluster. Apache Mesos can also perform push-based resource management but Spark can accept or reject the resources offered by mesos.

At runtime, Spark engages drives and workers to complete a task. While drivers define and invoke actions on RDDs, workers work toward storing RDD partitions and perform transformations. Figure 4-13 shows the engagement of driver and executors when a spark job is submitted. Let us go through the chain of processes during this event.

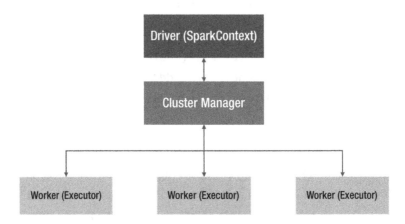

Figure 4-13. *Spark driver to executor communication*

Once a spark application is submitted to the cluster –

1. A spark driver is initiated and takes end to end responsibility of application execution. It is a JVM process which acts as a master node to host SparkContext for an application. It is responsible for breaking down an application into tasks and schedule them to run on executors, when available. If the driver dies, the process dies for some reason, executors also shut down.

    a.   SparkContext is the Spark application master which is responsible for setting up internal services and establish a connection to Spark execution environment.

    b.   It is used to create RDDs, accumulators, broadcast variables, access spark services, and run tasks.

    c.   RDDs reside within a logical boundary of SparkContext and is differentiable by its unique ID in a SparkContext.

2.   Spark driver invokes application's main () method.

3.   Driver coordinates with cluster manager for resources or executors to run tasks on worker nodes.

4.   Cluster manager launches executors, identified by an ID and the host. Executors are worker slaves that run in a JVM process on spark compute nodes and perform serial execution of the tasks assigned to them by the driver.

    a.   Executors register themselves with the driver to establish communication and receive tasks for execution.

    b.   Executors emit a heartbeat along with task metrics to the driver.

5.   Spark driver executes the application and assigns tasks to executors.

6.   Executors work on the assigned tasks and save results

7. Executors are terminated after main () method finishes or Spark driver runs `SparkContext.stop ()`. Thereafter, the resources are released back to the cluster manager.

RDD Composition

An RDD interface consists of –

1. Set of Partitions – Maintains the number of splits created for given RDD over the datasets.

2. List of dependencies – Maintains the dependency of the given RDD with the parent RDDs.

    a. Dependencies could be either *narrow* or *wide*. A dependency where each partition in the parent RDD is used by a maximum of one child RDD is termed as narrow dependency. If there are multiple child RDDs dependent on a single parent partition, the dependency is wide.

    b. Wide dependency can shuffle the data across nodes while narrow dependency follows pipelined execution.

3. GetPartitions – Compute the number of partitions on given RDD.

4. Data Locality – To avoid high shuffling of data it creates resultant datasets on the preferred locations.

5. Optional Partition information for creating key value paired RDD for accessing specific data at ease.

Datasets and DataFrames

A new interface was added in Spark 1.6 that provides the benefits of RDDs (strong typing and ability to use powerful lambda functions) and Spark SQL's optimized execution engine. Dataset can be constructed by JVM objects and then manipulated using Spark actions (map, flatMap, filter, etc). Currently Java and Scala support dataset whereas Python doesn't support it yet, but due to its dynamic nature it already provides many benefits of the Dataset API.

A DataFrame is a dataset structured into named columns. It is an enhanced feature to give a feel of tables in a relational database over structured data, files, tables in Hive, external databases, or existing RDDs, but with a richer optimization under the hood. DataFrame is available in Scala, Java, Python, and R with different types of representations.

Let us run through a small demo to create and perform basic functions with a DataFrame in Spark –

1. Create a Spark Session – to get all functionalities easily.

```
import org.apache.spark.sql.SparkSession
val spark = SparkSession
 .builder()
 .appName("Spark SQL basic example")
.config("spark.some.config.option",
"some-value")
 .getOrCreate()
// For implicit conversions like converting
RDDs to DataFrames
import spark.implicits._
```

2. In a spark session, create a DF from a file, existing RDD, Hive tables, etc.

```
val df = spark.read.json("examples/src/main/
resources/people.json")
// Displays the content of the DataFrame to
stdout
df.show()
// +----+-------+
// | age|   name|
// +----+-------+
// |null|  James|
```

3. DataFrame Operations

    i.   Print Schema of the DataFrame

```
// Print the schema in a tree format
df.printSchema()
// root
// |-- age: long (nullable = true)
// |-- name: string (nullable = true)
```

    ii. Select specific column

```
// Select only the "name" column
df.select("name").show()
// +-------+
// |   name|
// +-------+
// |  James|
// |    Ben|
```

iii. Select all columns, increment age column by 1.

```
// Select everybody, but increment the age
by 1
df.select($"name", $"age" + 1).show()
// +-------+---------+
// |   name|(age + 1)|
// +-------+---------+
// |  James|       22|
// |    Ben|       31|
```

iv. Filter the dataframe with age >25.

```
// Select people older than 21
df.filter($"age" > 21).show()
// +---+----+--+
// |age|   name|
// +---+----+--+
// | 31|    Ben|
// +---+----+--+
```

v. Count people by age.

```
// Count people by age
df.groupBy("age").count().show()
// +----+-----+
// | age|count|
// +----+-----+
// |  22|    1|
// |  31|    1|
// +----+-----+
```

4. Loading Hive tables

```
val table1 = spark.sql("[db_name].[table_name]")
table1: org.apache.spark.sql.DataFrame = [col1:
datatype, col2: datatype... 71 more fields]
```

5. Running SQL queries

```
// Register the DataFrame as a SQL temporary view
df.createOrReplaceTempView("people")
val sqlDF = spark.sql("SELECT * FROM people")
sqlDF.show()
// +----+-------+
// | age|   name|
// +----+-------+
// |  21|  James|
// |  31|    Ben|
// +----+-------+
```

Bucketing, Sorting, and Partitioning

In a file based data source, it is also possible to bucket and sort or partition the output.

```
peopleDF.write.bucketBy(42, "name").sortBy("age").
saveAsTable("people_bucketed")
usersDF.write.partitionBy("favorite_color").format("parquet").
save("namesPartByColor.parquet")
```

Deployment Modes of Spark Application

A Spark application has quite a few options for deployment. It can be either submitted in local mode, standalone mode, or cluster mode. The mode must be specified in the master parameter of a SparkContext.

1. Local Mode – In local mode, all spark job related tasks run in the same JVM.

    a. local uses 1 thread only.

    b. local[n] uses n threads.

    c. local[*] uses as many threads as the number of processors available to the Java virtual machine. For example,

    ```
    # Run application locally on 8 cores
    ./bin/spark-submit --class org.apache.
    spark.examples.SparkPi --master local[8]
    /path/to/examples.jar
    ```

2. Standalone Mode – In standalone cluster mode, spark allocates resources based on cores. By default, a spark application will try to consume all the cores in a cluster. The standalone cluster mode is subject to a constraint that only one executor can be allocated on each worker per application. In this mode, users can define containers for the worker and Spark master to run in your machine.

    ```
    # Run on a Spark standalone cluster in client
    deploy mode
    ./bin/spark-submit --class org.apache.
    spark.examples.SparkPi --master
    spark://207.184.161.138:7077 --executor-memory
    20G --total-executor-cores 100 /path/to/
    examples.jar
    ```

```
# Run on a Spark standalone cluster in cluster
deploy mode with supervision
./bin/spark-submit --class org.apache.spark.
examples.SparkPi --master spark:
//207.184.161.138:7077 --deploy-mode cluster
--supervise --executor-memory 20G --total-
executor-cores 100 /path/to/examples.jar
```

3. Cluster Mode – A Spark application on YARN can
    be launched either in cluster mode or client mode.
    In cluster mode, the Spark driver becomes the part
    of application master, which is entirely managed
    by YARN. In client mode, the Spark driver runs as a
    client process which coordinates with application
    master to manage resources for the Spark application.

```
# Run on a YARN cluster
./bin/spark-submit --class org.apache.spark.
examples.SparkPi --master yarn --deploy-
mode cluster # can be client for client
mode  --executor-memory 20G --num-executors 50
/path/to/examples.jar
```

A Spark application can also be deployed using mesos, but due to
its ability to offer Spark driver to choose resources, it has been restricted
to a push-based resource allocation mechanism. For mesos, the master
parameter will be mesos: //host:port.

Design Considerations

Below are the key design considerations for deploying spark applications
on a Hadoop cluster.

1. Oversubscribing cores is a useful way in Spark to avoid time in context switching in assigning tasks. There are internal threads in JVM that will run shuffle, GC operations which uses the same cores assigned to the application.

2. If the Resource Manager(RM) crashes application will not get affected until any running container is crashed or in case of dynamic allocation it tries to negotiate resources to RM.

3. RM has as scheduler which has the information about each application running on cluster, it knows which node is assigned as Application Master (AM) for the respective application, if any AM is crashed RM will assign to other Node Manager (NM) by round robin scheduling.

4. Data locality comes into the picture for big clusters, if the application is started on fewer executors.

5. Running in client mode on cluster environment has a drawback. It will terminate the application if the remote client gets disconnected from the network. It becomes a single point of failure.

6. Running in cluster mode minimizes network latency between the drivers and the executors, the chance of network disconnection between "driver" and "Spark infrastructure" reduces. Since they reside in the same infrastructure. It also, reduces the chance of job failure.

7. If the executor gets crashed or lost then Spark's DAGScheduler and its lower level cluster manager implementation (Standalone, YARN, or Mesos) will notice a task failed and will take care of rescheduling the said task as part of the overall stages executed.

Caching and Persistence of an RDD in Spark

Re-computation of RDD after each action call affects the performance of the program based on consumption of resources and time. To enhance this process Spark came up with two optimization techniques *Cache* and *Persists*. The aim of both of the methods is to store the dataset into memory/disks temporarily and to reuse the results iteratively over multiple computations in multistage applications. It provides multiple ways of storing and replication of data. Below are the scenarios for when caching should be switched on for usage –

1. In Standalone spark applications

2. In Machine Learning applications

3. In expensive RDD computations in a resource constraint environment, caching helps reducing the cost of recovery, if any executor gets failed.

RDD Cache is used to speed up the apps that access the same RDD several times. With cache you use only the default storage level MEMORY_ONLY.

RDD Persist provides multiple options to store dataset either on memory or on disk. The difference between cache and persists is purely syntactic. The persist method takes place in respective storage levels including:

- MEMORY_ONLY (default level) – It stores the data as a deserialized object, if there is insufficient memory some of the data partitions may not be cached, that uncached data will be computed next time when we need it.

- MEMORY_AND_DISK – RDD stored as deserialized data objects, if RDD may not fit in the memory cluster, it stores the remaining part on the disk.

- MEMORY_ONLY_SER – RDD are stored as serialized Java objects in memory. Serialized object means one-byte array per partition. This is much more space efficient, which saves memory.

- MEMORY_AND_DISK_SER – Similar to MEMORY_ ONLY_SER but it saves the leftover part in the disk.

- DISK_ONLY – Stores the RDD partition only on Disk.

- MEMORY_ONLY_2, MEMORY_AND_DISK_2 – These two levels work the same as the above two, but these two replicate each partition on cluster nodes.

Spark implicitly monitors the cache usage per node and purges old data based on the least recently used approach. An RDD can be manually removed using RDD.unpersist () method.

RDD Shared Variables

If a spark operation requires a function for execution on the cluster, separate copies of function variables are created and copied on each machine. Any changes to these variables do not propagate back to the spark driver. To overcome this situation, spark provides two categories of shared variables, namely, broadcast variables and accumulators.

1. Broadcast variables – instead of creating a copy
    of function variables, it will keep a read-only
    copy cached on each worker nodes. Broadcasting
    becomes an operation-in-demand when the
    common data needed by the tasks on different
    nodes can be shared between stages. Efficient
    broadcasting algorithms help in optimizing
    internode communications.

2. Accumulators – Accumulators are the variables
    that result from an associative or commutative
    operation. Spark supports named or unnamed
    accumulators of numeric value types.

SQL on Hadoop

Imagine you have just migrated your relational data warehouse to a
Hadoop platform. The business users who were comfortable with SQL
queries, are now finding it difficult to perform routine data checks. In
another instance, data analysts want to run ad-hoc queries for interactive
analysis, but feel restricted by the transition curve from a *SQListic*
approach to a developer one. So, what's the solution?

Having learned the motivation behind Hive in this chapter and its
adoption and makeover from traditional MapReduce, the fact that classical
database community is still going strong cannot be understated. There are
two challenges while accessing the data in a lake, the language and data
scale. While a 'language' is required for interactive data exploration, the
ability to cope with peta-scale data sets makes a lot of difference.

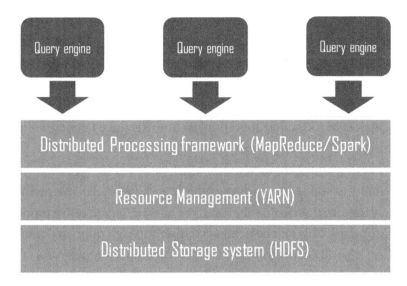

Figure 4-14. *SQL on Hadoop layout in a data lake*

SQL on Hadoop provides a platform to enable database community to play around with data in Hadoop without knowing much about MapReduce and cluster computing. By no means, does it replaces data warehouse with a "low-cost" excuse or mimic a database processing engine, rather it provides an abstraction over HDFS and YARN to empower data analysts. The figure below shows the usage patterns of SQL on Hadoop. Depending on the usage pattern, the architectural guidelines differ for the platforms and their usage.

Analytical	Batched	Transactional
Interactive ad-hoc analysis	ETL operations	ACID support
Acceptable latency	Select and DMLs	Select and DMLs
Apache Drill, HAWQ, Splice	Acceptable latency	Acceptable latency
Oracle BigData SQL, Presto	Hive, SparkSQL	Apache Trafodion, Splice, Phoenix

Figure 4-15. *Operation modes of a SQL on Hadoop framework*

The key aspect of SQL on Hadoop is the engine responsible for parsing the query and processing it for data extraction. An ideal SQL on Hadoop engine should be a distributed one and must possess the ability to scale out seamlessly. While the data movement in and out of a data lake should be minimal, it must overcome the latency and concurrency bottlenecks. Unlike a typical MapReduce operation, it should reduce latency while maximizing concurrency.

Another important characteristic of SQL on the Hadoop framework is extensibility or the ability to query other data sources. An interactive query may require data from other sources as well to generate a report or feed a dashboard. This can be achieved either through query federation or franchising, or pull agents to stage data at a centralized repository.

In the next few sections, we are going to discuss how to position Presto, Oracle Big Data SQL, or Cloudera Impala in a data lake ecosystem.

Presto

Presto is a distributed query engine which was started by Facebook in 2012 to enable faster and scalable queries on very large-scale datasets. Being developed in java, it supports ANSI SQL semantics including complex joins, aggregations, and analytical functions. Presto was designed to suffice low latency data analysis and data warehousing requirements.

How does a Presto engine work? Unlike Hive, presto doesn't use MapReduce for query execution. It implements a standard MPP query processing engine comprising of coordinators, a discovery service, workers, and connector plugins. Discovery service, as the name suggests, acts as an engine coordinator which receives heads-up from the participating nodes (or servers) at the time of startup. *Presto coordinator* is responsible for statement parsing, plan generation, and managing workers. A client submits a SQL statement to the coordinator via REST API, who works with the connector plugin to receive table metadata and split details for building the query plan. Once the query plan is generated, it breaks them into a series of stages and further into tasks, and distributes them over to the worker nodes. *Workers* work with the connector plugin for the execution of tasks within memory. They fetch data splits from the connector plugin, and run parallel drivers for processing as per the operation. Once completed, task output is transferred over to another worker interchangeably. Workers follow pipelined aggregation, i.e., each worker performs aggregation and transfers task output to the next worker. Once the aggregated data reaches the last worker, it is sent back to the client via coordinator.

Connectors are nothing but the drivers to access meta information and data from the data stores. Presto can access data not just from HDFS, but also other disparate data sources. Connectors are available for Hive, Cassandra, and Postgres, and a few more data stores. Each connector has access to its metadata. For example, Hive connector can access Hive metastore. MySQL connector will have access to its own catalog information. Figure 4-16 shows the flow of query execution from the client submission until the result set.

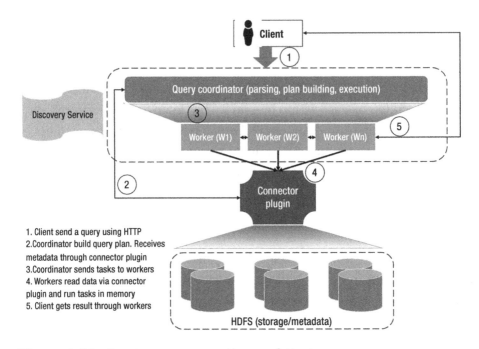

Figure 4-16. *Presto query execution architecture*

Presto Statement Execution Model

Figure 4-17 shows the steps followed during execution of a SQL statement.

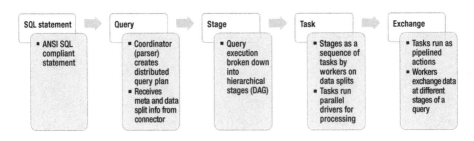

Figure 4-17. *Presto statement execution steps*

Consider a scenario where a SALES dataset from Hive needs to be joined with the PRODUCTS dataset in MySQL. Let us check how the catalog entries look like –

```
$ cat etc/catalog/mysql.properties
connector.name=mysql
connection-url=jdbc:mysql://localhost:3306
connection-user=root
connection-password=*****

$ cat etc/catalog/hive.properties
connector.name=hadoop2
hive.metastore.uri=thrift://localhost:8180
```

The below Presto query joins two datasets to count the point in sales per product. This query is self-explanatory and follows ANSI SQL code practices.

```
SELECT
    p.prod_code
    ,COUNT(*) AS num_sales
FROM hive.sales.sales_ds s
    JOIN mysql.products.prod_desc p
ON s.prodid = u.prodid
GROUP BY p.prod_code
```

Presto – Design Considerations

1. Presto is not a database, but provides SQL capability for data processing through a pipelined execution model.

2. A SQL statement is converted into query, stages, tasks, and splits running in parallel.

    a.   Query plan is dynamically compiled into an optimized m-code

3. Performance management

    a.   Presto uses direct memory management.

    b.   Memory assigned to a worker must be a function of workload running on the node. If you run relatively complex queries, cluster might report latency issues.

    c.   Presto performs multiple in-memory operations throughout the query execution and relies on network for transfer of data. It doesn't take care of implicit fault-tolerance.

    d.   Use ANSI SQL best practices for query tuning

        i.   A CTAS command works faster than SELECT *.

       ii.   To optimize joins, reduce the data surface area by reducing the size of data sets being joined. Note that Presto runs broadcast joins by default, which means a fact table can be distributed, while a dimension table can be copied over to the worker nodes for processing.

      iii.   Inline views can be accommodated in a WITH clause subquery.

      iv.   Single node operations could be memory intensive like DISTINCT count, ordering, and stitching data using UNION must be avoided. Use APPROX_DISTINCT in place of DISTINCT.

     v.   Multiple LIKE can be replaced with REGEXP_LIKE.

    vi.   More than 1000 AND/OR in a query may result in a compiler exception.

4. Ensure the version compatibility between connector and Hadoop version while creating catalog properties. For example, a Hive connector should be able to access Hive metastore as well as fetch data splits of a table from the cluster.

5. Catalog – A query in presto is run against one of the registered catalogs. A data source catalog must be registered by creating a catalog properties file in /etc/catalog/ directory. A connector can be mounted using connector.name parameter, which indicates that the catalog manager will create a connector using this catalog.

    a.   Create multiple catalogs of similar data source type with different configurations. For example, two Hive cluster catalogs will have the same connector.name i.e. Hive, but different configuration.

```
connector.name=hive-cdh4
hive.metastore.uri=thrift://[master]:10000

connector.name=hive-hadoop1
hive.metastore.uri=thrift://[master]:10020
```

6. Monitoring – Presto provides an interactive web interface, Presto-Admin, for the management and monitoring of queries.

a. Check status of all presto nodes

    `presto-admin server status`

b. Presto clients – Airbnb designed a query execution tool, Airpal for easy query writing and object browsing.

c. The presto page `https://prestodb.io/resources.html` collates all information regarding management tools, clients, drivers, plugins, and libraries. It is maintained by Facebook.

d. For uninterrupted operations, deploy a periodic monitoring script to take proactive actions in case it detects a configuration issue, and sends out alerts and notifications.

7. Configuration properties

    a. Cluster

        i.   `task.info-refresh-max-wait` – optimize coordinator workload

        ii. `task.max-worker-threads` – split and assign process to worker nodes

        iii. `distributed-joins-enabled` – Enable distributed hash joins

        iv. `query.max-memory` and `query.max-memory-per-node` – max allocation of distributed memory at engine and node level

        v.   `discovery-server.enabled` and `discovery.uri` – Discovery service URI specification.

b. Coordinator

    i. `query.initial-hash-partitions` – maximum hash buckets of an aggregation

    ii. `node-scheduler.min-candidates` – max parallel workers to run a stage

    iii. `node-scheduler.include-coordinator` – should coordinator run tasks?

    iv. `query.schedule-split-batch-size` – Stage to task split size

c. Worker

    i. `task.cpu-timer-enabled` – detailed statistic collection

    ii. `task.max-memory` – memory restriction for CPU intensive operations like joins and sorts.

    iii. `task.shard.max-threads` – worker threads to run active splits. Set it as (4*CPU cores).

d. Performance

    i. Hive.empty-bucketed-partitions. enabled=true

    ii. Hive.bucket-execution=true

    iii. Hive.assume-canonical-partition-keys=true

    iv. Hive.multi-file-bucketing.enabled=true

    v. Hive.immutable-partitions=true

8. Hive to Presto migration

    a. Note the difference between HiveSQL and ANSI SQL

    b. Follow ANSI SQL guidelines while using arrays, quoted identifiers, and SQL operators. For example,

        i. Use UNNEST in Presto to expand arrays, instead of LATERAL VIEW in hive

        ii. '2APR2014' is not valid in Presto, but "2APR2014" is

    c. For troubleshooting, focus on operator usages in query predicates. For example, [column = NULL] is different from [column is NULL]

    d. The VALUES clause could be effective in testing Presto query. It helps in preparing a sample data set for testing

Oracle Big Data SQL

Oracle Big Data SQL is a SQL on Hadoop solution from Oracle that allows issuing a query against Hadoop, Hive, Oracle NoSQL, and HBase from Oracle database. Put it simply, it brings non-relational data under one roof of Oracle system catalog. As a result, Hive table access benefits from most of the smart features of Oracle database like query offloading, smart scan on Hadoop, network resource management, data redaction, and other Oracle advanced security features, without any compromise. Existing users of Oracle advanced analytics and security can expand their data coverage by switching on Big Data SQL and all their running models will start considering data from Hadoop data lake.

Figure 4-18 shows the benefits of Oracle Big Data SQL product.

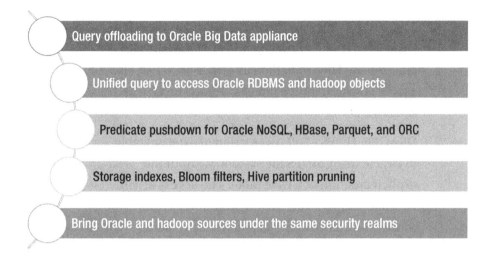

Figure 4-18. *Oracle Big Data SQL benefits*

How Oracle Big Data SQL works? Well, contrary to conventional techniques of data federation, Oracle Big Data SQL uses the *query franchising* approach for unified processing across relational and Hadoop source. Query franchise promotes driver based query execution, rather than dispatching subqueries to different data systems for their native processing engines to work upon. Drivers or agents are compute agents on different systems (Hadoop data nodes) which assist the execution of a task in Big Data SQL. What this franchisee model achieves is unified resource management and effective query planning as Oracle optimizer understands data location and structure, spanning across multiple systems. The agent or Oracle Big Data SQL cell optimizes query execution through smart scan and storage indexes.

Big Data SQL follows external table mechanism to access table metadata via Hive metastore. Subsequently, it also uses the underlying Hadoop APIs to access data from HDFS. The driver, also called storage handlers, must be specified while creating external table in Oracle database. To access Hadoop cluster, there are two access drivers available in Oracle –

1. ORACLE_HIVE – Allows creation of Oracle external tables using Hive metadata information. It can also access HBase, if a Hive table is defined on HBase store.

2. ORACLE_HDFS – Allows creation of external table in Oracle using HDFS file system, without the explicit creation of a Hive table. The access driver implicitly imitates the Hive syntax and assigns a schema structure to the data from filesystem.

The below CREATE TABLE script creates an external table for Hive table social_cmt_hv.

```
CREATE TABLE ratings_db_table (
    col0 VARCHAR2(4000),
    col1 VARCHAR2(4000),
    col2 VARCHAR2(4000)
)
ORGANIZATION EXTERNAL
    (TYPE ORACLE_HIVE DEFAULT DIRECTORY DEFAULT_DIR
    ACCESS PARAMETERS
      (
        com.oracle.bigdata.cluster=hadoop
        com.oracle.bigdata.tablename=default.social_cmt_hv
      )
    ) PARALLEL 2 REJECT LIMIT UNLIMITED
```

The ACCESS_PARAMETERS clause can be used to specify a replacement action in place of its default behavior. For example,

- if table names are different in Oracle and Hive, specify Hive table name in `com.oracle.bigdata.tablename`

- if column names are different, specify column mapping in `com.oracle.bigdata.colmap`

Design Considerations

1. Oracle data dictionaries [USER] DBA_HIVE_TABLES contain Hive table metadata details.

2. Oracle Big Data SQL is a separately licensed product and compatible with Oracle database version 12.1.0.2.0 and higher.

3. Oracle SQL Developer 4.0.3 comes with the capability of connecting to Hive metastore.

4. Functions that can be offloaded to Hadoop can be queried from v$sqlfn_metadata dictionary.

5. You can generate DDL of an external table to access a Hive table in Oracle database using `dbms_hadoop.create_extddl_for_hive` subprogram

6. Big Data SQL 3.0 and onwards can be installed on commodity hardware with cloudera CDH 5.5 or hortonworks HDP 2.3 distribution of Hadoop.

   a. Network optimization due to infiniband fabric in engineered systems cannot be leveraged.

7. Smart scan – The full scans of Oracle external tables that use access drivers for Hive or HDFS, are optimized through smart scan capabilities by the BigData SQL cell running on Hadoop data nodes. Smart scan is a feature of Oracle engineered systems that optimizes query execution by pushing down processing to the storage layer.

8. Storage indexes – Oracle Big Data SQL cell maintains storage indexes for the data distribution over HDFS. Storage index helps in eliminating the scans of those disk sectors which don't have the required data blocks. Smart scan and storage indexes complement each other for query optimization. For effective use of storage indexes, use frequently used column to sort the query. A query using equality (=), non-equality (<, >, <>), null (IS NULL, IS NOT NULL), or less than/greater than (<=, >=) operators are benefitted from storage indexes

9. Queries against ORC and parquet file format are further benefitted by the stripe indexes within the file structures.

10. Oracle database 12c has developed native capability to store and parse JSON data. Pushdown of CLOB processing is possible with Oracle Big Data SQL.

Conclusion

Data processing has been through multiple innovations in the recent times. Modern day cutting edge frameworks like Spark, presto, and SQL on Hadoop technologies have eased the life of data practitioners who are transitioning from relational world to unstructured space. What remains common within newcomers is extensibility and integrity.

This chapter tried to discuss the data processing techniques in data lake. Many of you might be working on these platforms but understanding the design considerations and impactful factors is the key to a successful implementation. Within the scope of this book, we discussed classical MapReduce, Hive, Pig, Spark, and presto. However, there could be multiple other frameworks and designs like cloudera impala or open source products, which serve a similar purpose.

In the next chapter, we are going to introduce data governance along with one of its key accountables: data retention and archival strategy.

CHAPTER 5

Data Archiving Strategies in Data Lakes

'Data is like garbage. You'd better know what you are going to do with it before you collect it'

—Mark Twain, an American writer and entrepreneur

The linearly growing data lake sophistication trend has empowered the rise of data analytics from descriptive to predictive, and further to prescriptive. The strategies that drive meaty business outcomes, rely heavily on data initiatives that offer quality and relevance. An enterprise data lake, being the mainstay of modern cognitive data analytics, banks upon a body that guards its lifecycle through the stages of transformation and consumption. How often do you see an analyst questioning data sufficiency for a data model? How often does security analysts mark risk zone for data lake applications to measure their vulnerabilities? Here comes the role of data governance – a key pillar to overall data strategy in an organization.

© Saurabh Gupta, Venkata Giri 2018
S. Gupta and V. Giri, *Practical Enterprise Data Lake Insights*,
https://doi.org/10.1007/978-1-4842-3522-5_5

In this chapter, we'll draw our focus on one of the deliverables of data governance, namely, data archival strategy. We'll understand why data archival is crucial for an enterprise and in a data lake ecosystem. Later, we'll explore design considerations that potentially drive data lifecycle management strategy in an organization.

The Act of Data Governance

Data governance is the "council" who is aware of the mission and vision of business outcomes and is entrusted with the task of aligning enterprise data initiatives with those outcomes. Data governance owns the entire gamut of data lake and formulates the vision, strategy, and framework within the organization. As per the Forbes research in 2016,[1] 78% of leading BI Executives recognize the importance of governance in BI, while 65% accept that governance offers useful means to empower end-users to uncover new insights. Mike Saliter, VP, Global Industry Solutions at Qlik, realizes the importance of governance council and explains:

> *Governance requires a really fine balance - governing to the point where consistency is assured, but flexibility remains. There is no perfect formula, but finding the right governance level within your organization's culture is a critical component to making the most of BI opportunities.*

The council, headed by a chief data officer (CDO), folds in key stakeholders from data teams to form a data leadership guild. Data leadership guild takes the decision on:

---

[1] https://www.forbes.com/sites/forbespr/2016/10/24/strong-data-governance-enables-business-intelligence-success-says-new-forbes-insights-study/#6832a44f582d

1. Streamline data approach across organization – lay down best practices, models, and process flow for data engineering, ingestion, and visualization

2. Data privacy and compliance – Verify if data sources are compliant.

3. Data inventory practices

    a. Data acquisition linked with business deliverables – Procure new data sources to transform data lake into a data asset. Advise curation layer as information libraries to hold *prepared* data for business analytics teams.

    b. Data retention – devise evaluation strategy to retain or suspend data system from data lake

    c. Data security – implement security fencing around data and platform to mitigate risks associated with data in company and public networks

        i. Data masking

        ii. User authentication and authorization

4. Infrastructure planning

    a. Storage – Implement information lifecycle management (ILM) policies to archive data

    b. Capacity planning – Depending on the user base and business forecasting, plan the infrastructure to ensure business continuity

Figure 5-1 shows a typical data governance council organizational structure.

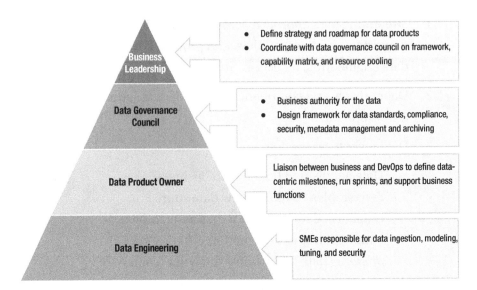

Figure 5-1. *Data governance organizational structure*

Data lake vs. Data swamp

Data lake follows an ordered way of bringing in the data. It strikes "one-for-all" approach to lay down data ingestion strategy, data organization, and architecture. Data governance council reviews the data patterns and practices from time to time and ensures that data lake follows secure, steady and sustainable approaches. A methodological approach to soak in all complexities (data sources, types, conversion) under a common layer is vital for successful data lake operations.

Data swamp, on the other hand, presents the devil side of a lake. A data lake in a state of anarchy is nothing but turns into a data swamp. It lacks stable data governance practices, lacks metadata management, and plays weak on ingestion framework. Uncontrolled and untracked access to source data may produce duplicate copies of data and impose pressure on storage systems.

Figure 5-2 shows key differentiators between data lake and data swamp.

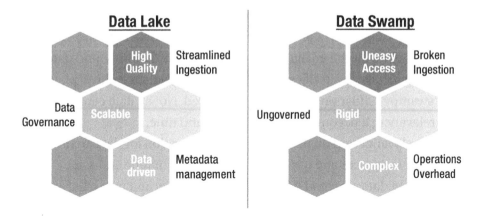

Figure 5-2. *Comparison between data lake and data swamp*

To prevent data swamp situation, enterprises can adopt best practices listed as below.

1. Advocate data governance – Data governance ensures appropriate sponsorship of all resident data and tags data initiatives with visible business outcomes. Looks after security aspects and participates in capacity planning of data lake.

2. Build and maintain metadata – Metadata management should be encouraged to ease data access and support qualitative data search.

3. Prioritize DevOps charter – DevOps keeps the ingestion pipeline intact and shields data complexities under generic framework.

Introduction to Data Archival

Data archival refers to the techniques to retain infrequently used data at a lower cost. For huge data sets, massive storage is required. For efficient data access, smart storage is warranted. Faster and smarter storage systems may increase the operational expenses and thus, the total cost of ownership (TCO). Data council keeps an eye on the data relevance and draws a timeline for the analytical models to enable a time-bound data visibility. Data lake, by itself, is principled to act as a data archival platform but a "swampy" situation needs to be prevented. Not all data can be of equal relevance. It becomes dormant by age and time.

Data relevance and quality drive strong motivation for data archival strategy. Data sources and quality of data evolve with time and age. A 10-year-old data set could be of suboptimal quality and may offer no material when fed to a learning based analytical model. Keep in mind that easy and rapid data access is one of the highlights of data lake flyers. Having said that, dormant data sets may still reside in the lake for regulatory and compliance reasons, but at a cost-effective storage. The ability to manage data lake lifecycle depending upon age and relevance of data draw a thin line between a costly data lake and the one that is "spot on" economical.

As data grows on the scale of terabytes and petabytes on tier-1 storage systems, it may impact primary applications by slowing down the data access. Obviously, underlying storage systems face cost and space pressure due to this data growth. In addition, data operations like backup, restore, analyze, or clone run longer due to sheer volume of data. Such scenarios directly impact IT budget and bring down the return on investment (ROI). Data archival practice helps in making room for new data. Not only archival strategy optimizes load on primary storage system, but it may also result in improved performance. Performance optimization is attained by reducing the surface area of active data volume in lake and not by deploying any tuning techniques. Figure 5-3 shows a flyer of data archival benefits.

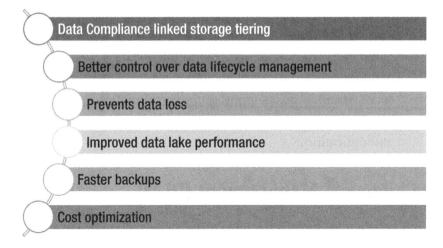

Figure 5-3. *Benefits of data archival strategy*

Cloud based storage could be a big relief for enterprises. Public cloud vendors claim to offer elastic storage services at reasonably low rates. Although it depends and differs case by case, but cloud storage and archival services assure better returns on affordable capital because of lower operational expenses (no site space required and minimal operations). Let us quickly go through few factors which an architect must consult before positioning cloud archival strategy –

1. Recovery turn-around time – Cloud service model must support quick data recovery. Recovery approach must be simple and interactive

2. Data lifecycle integrated management – Archival service must have the capability to be integrated with other cloud storage classes. A unified lifecycle policy can be defined to implement storage tiering using primary as well as archival storage class.

3. DLM Compliance and regulations – Archival service must align with the DLM considerations like data awareness, data search, and cost optimization.

4. Security – One of the key parameter of cloud versus on-premise debate. Security is largely subjected to data governance council's definition of data confidentiality and compliance. However, a cloud service vendor must support security techniques like data encryption, access management, and authentication.

Data Lifecycle Management (DLM)

Data lifecycle management is the ability to classify data by its age and business relevance and define policies to move data across the storage tiers. Factors that drive an effective data lifecycle management are:

1. Data awareness – Very important to have sound understanding of business service level agreements and data model awareness. It helps in determining what the data is, opposed to where it is situated. Data awareness helps in learning legality and sensitivity of data.

2. Data retention, transition, and expiration policies – Storage and compression tiering can be implemented through lifecycle rule definitions. The rules may define ageing policies, which when triggered, may execute appropriate action like data movement or deletion. Figure 5-4 shows the circle of life of data in a lake.

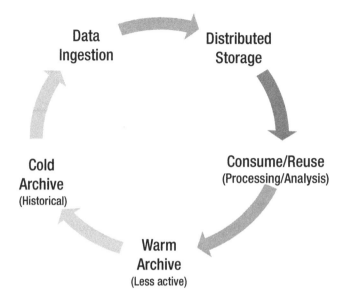

Figure 5-4. *Data Lifecycle Management flow in a data lake*

3. Familiarity with the data sources for business-
    relevant classification – Unless and until one
    understands the significance of data, he/she will not
    be able to determine the impact of data archival. For
    instance, retail transactions older than two years
    can be archived but retail contractors can never be
    archived.

    a. It is not a good approach to put hard timeline
        for all objects in data lake. It could have an
        adverse impact on the business relevance of
        data. For example, a customer profile created
        prior to the hard timeline may no longer be
        visible in the lake and thus to the models that
        implement consumer analytics.

    b.   Archive data as per the business dependency – Viable approach to archive data after reading through all dependencies. Customer transactions before a stipulated date may be archived. However, customer data will still be active and visible to the system.

4.   Frequency of data access – Helps in defining ageing rules for data.

5.   Effective governance and compliance – Data governance has the full view of data and their storage classes.

DLM policy actions

DLM policies can be defined to act based on business needs and data relevance. Data archives can be online or offline. Data can either be moved to a connected low-cost storage or purged to a storage that is disconnected from the active data lake. While most enterprises give preference to online data archives, offline archives come to play in two situations:

1.   Source system retires or gets decommissioned and its footprints in the lake needs to be archived.

2.   Source system goes through a transformation via digitization initiative and legacy raw data sets need to be archived.

For online data archives, DLM policies can implement storage tiering and compression tiering. A storage tier refers to the bucket that holds data of a defined classification and abides by a DLM policy. Storage tiers may have different underlying storage and related attributes like compression mode and disk speed.

DLM strategies

Data lifecycle management should rather be a stepwise approach for defining archival policies than an instantaneous action. Below are the steps to be followed to design a strategy:

1. Data prioritization – Selection and classification of data should be the first exercise of DLM strategy. Data can be zoned out based on its business criticality and dependencies it shares upon other entities. The categories and parameters are fully dependent upon the data council regulations. However, standard data catalog can be build be classification of data under below categories:

    a. Master data – data of high and consistent business importance needs to be preserved unless and until data governance approves for archival. Master data can be sensitive information like Personal Identifiable Information (PII) data (SSN, customer, employee, patient) or company's internal data (products, vendors, suppliers, deals, and agreements). For healthcare industry, HIPAA privacy rule specifies the number of years for which a patient record must be retained. Baking industry mixes multiple parameters like last transaction date and balance to classify an account as dormant.

     b.   Transactional data – data that can be archived without any issues. Policies can be defined to archive daily transactional data after it attains certain age.

         i.   Machine or sensor generated data older than two years

        ii.   Location data older than 30 months

      iii.   Bank transactions older than five years

      iv.   Educational data older than one year

     c.   Social data – social data can be archived more often than other categories because most of the analytical models that consume social data acquire intelligence through continuous and progressive learning. And as these models mature, they marinate raw data streams with their insights, only to enrich them. Therefore, data from social media has the tendency to grow dormant faster. Monthly archival policy to move web logs to a secondary storage may suffice the storage requirements.

2.   Data age by classification – for each classification of data, the age parameter may differ by the data's relevance or governmental norms. Setting the time dimension per classification would establish the link between archival policy and data classification.

3.   Define archival policy – archival policy can be declared for each data source at storage level. Each policy carries a name, rule, and an action. Policies can be triggered implicitly or manually.

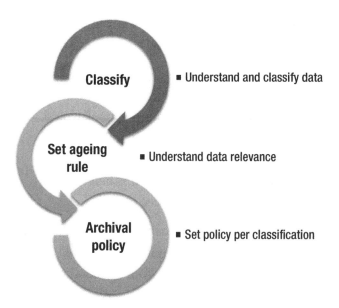

Figure 5-5. *Data Archival strategy foundation*

DLM design considerations

Let us check out key design considerations that can be tactical while implementing a comprehensive data archival strategy (Figure 5-5).

1. Data backups as data archives – Avoid using data backups as data archives. It has been a long running debate that periodic data backups can be treated as archives. However, the reality is backups and archives are two different worlds. Backups are like data fillers when lost, while archives are data finders.

   a. Backups are used if when you are restoring or cloning a system or during disaster recovery. Archives enable data discovery at the cost of performance.

b. Data backups are disconnected from the active data while archives are well connected with the lake.

c. For example, if a customer needs bank transaction older than a quarter, bank would pull up from the archives. Should it be pulled from the backups in the absence of archives, one must restore all the backups before accessing the required piece of data.

2. Archive performance – Accessing data from the archives may give a suboptimal performance because data has to be pulled from a secondary storage.

a. Crucial factor while mapping data classification to a storage tier in a DLM policy. For an archive, performance becomes the function of disk speed of secondary storage, network bandwidth, and compression mode.

b. If archive stay in compressed format, it would further degrade the performance as it has to be decompressed before access.

3. Archiving unstructured data – for unstructured data, you can set up an archival hadoop cluster of secondary storage systems. Storage paradigm remains the same as in primary hadoop cluster. There are multiple commercial vendors who offer products to archive unstructured data to a file system.

4. Discover data dependencies – you must fully understand the relationships between different data types before implementing the archival plan. We saw the example of customer profiles in DLM section. There are a few more listed below.

    a. Master business entities should never be archived unless the entities are decommissioned or suspended

    b. Avoid fixed timelines for archiving – *"All data until year 2000 will be archived"* – this rule will sunset all master data fed before 2000. Is it expected? Data governance needs to take the call.

    c. Call data can be archived after 15

5. Hadoop archives (HAR) – A mapreduce based utility to archive multiple small files on HDFS cluster into an immutable file. It is designed to tackle small file problem in hadoop.

    a. hadoop archive –archiveName dummy.har / input/location /output/location

6. Cloud based archives – Public cloud vendors offer archival storage service at affordable rates. Cloud based storage not only serves as a replacement to tape mechanisms, but also offers better availability, enterprise security, easy management, and compliance.

    a. Durability – On-premise archiving provides complete control but could impost cost pressure on enterprises. In addition, archive durability, availability, and security are some of the pressing issues to be addressed.

b.   Security – Though it is difficult to draw consent from everyone on the topic like "cloud" security, it might be good to know how cloud vendors are tackling security. Cloud vendors claim to encrypt the data as soon as it comes. Keeping in mind the level of skill artillery that cloud companies possess and level of compliance they comply with, they possess encryption keys for their customers. However, it might not be a bad idea to encrypt data once before pushing to the cloud; so that one set of encryption level is entirely owned by the customer.

c.   Can be used to archive legacy data or unstructured data.

d.   Commercial vendors that offer archival services

   i.   Amazon Simple Storage Service and AWS Glacier

   ii.   Microsoft Azure Archival Storage

   iii.   Oracle Cloud Infrastructure Archive Storage Classic

7.   Tiering levels – distribution of data into cold, warm, and hot buckets

a.   Hot data refers to the most active one residing on primary storage. All data drive actions and analytical models access hot data from the lake.

b.   Warm data is the dormant one which can be selected at times but never really participates in a transaction.

   c.   Cold data is the one that exists just for regulatory and compliance purposes. It doesn't participate in data access or transactional exercises.

8.   Retrieval approach – for both on-premise as well as cloud based archival, make sure that one is aware of how to retrieve the archives when required.

Amazon S3 and Glacier storage classes

Amazon offers S3 storage containers for object storage with high durability and finds variety of fitment. It offers data management console for monitoring and lifecycle control, data protection through versioning and replication, event and alert notifications, and security controls. We'll not deep dive into Amazon S3 service to abide by the scope and purpose of the book. However, we'll focus on the storage classes of Amazon S3.

Amazon offers a low-cost archive storage service, known as AWS Glacier. It is used for deep archival of data which is infrequently accessed but retained for compliance. If you are looking for a long-term backup solution, Glacier might well fit your bill. Amazon Glacier provides three methods to access data, namely, expedited, standard, and batched service. Depending on the data access requirements, appropriate retrieval mode can be selected (Figure 5-6).

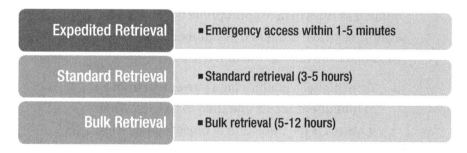

Figure 5-6. *AWS data retrieval policies*

From the security standpoint, all data within the glacier is encrypted. In addition, access control can be configured through AWS IAM service. Lifecyle features of Amazon S3 are listed below.

1. Tiered data management –Amazon S3 storage classes can be S3-Standard, S3-Standard-IA, or S3-Glacier.

    a. S3 Standard – It is the primary object level storage class for general purpose data storage. It serves data access requests which generally need low latency and high throughput. All hot and active data resides on this storage class.

    b. S3 Standard – Infrequent Access – The object level storage class is used for warm or less-frequently used data. Data access requests of low latency and manageable throughput can be served rapidly using Standard-IA storage class. Minimum storage duration is 30 days.

    c. Glacier – the object level storage class is used to archive cold data which is retained for compliance and regulatory purposes. Minimum storage duration is 90 days.

2. Lifecycle rules – Data lifecycle policies can be defined as rules to take appropriate action on data ageing. Data age is defined as number of days or a fixed date. The nature of action can be:

    a. Transitional – Enables data movement from S3 to infrequent access and then to Glacier. What it means is the transition of object storage class from S3 to STANDARD-IA or GLACIER.

b.   Expiration – Enables deletion of data in S3
     storage based on a date or age. Since AWS
     follows queueing approach to delete objects
     asynchronously, there could be a lag between
     expiration date and actual expiration.

The reason one should go for a cloud archival service lies in the
fact that total cost of ownership for cloud is way less than on-premise.
Compared to traditional archiving approaches like tape libraries, drives,
media, or specialized frameworks, AWS Glacier achieve intelligent
archiving at a reduced cost. Other major consideration would be
durability. AWS guarantees 99.9% of data durability along with regular
checks on fixity and automatic recovery during failures.

Design considerations

As a cloud architect, one must bear the following points in mind before
executing data archival strategy in an enterprise data lake:

1.   Storage classes can only be moved further. For
     example, an object in STANDARD class can
     only transition to STANDARD-IA or further to
     GLACIER. An object in STANDARD-IA cannot be set
     back to STANDARD. In such scenarios, archives can
     only be restored from Glacier archives.

2.   An object with age less than 30 days cannot
     transition to STANDARD-IA

3.   An object sized less than 128KB cannot be
     transitioned to STANDARD-IA

4. Based on the lifecycle action, pricing may differ.

    a. If you are deleting objects from STANDARD-IA storage class within 30 days, you might get charged for early deletion.

    b. Every archival as well as retrieval request in Amazon Glacier is chargeable.

5. Objects that are encrypted continues to be encrypted throughout the transitioning process.

6. Restoration from Glacier archives is a time-consuming process.

Figure 5-7 shows Amazon AWS storage class portfolio.

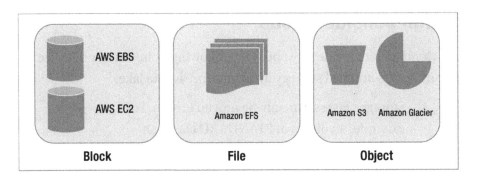

Figure 5-7. *Amazon AWS storage class portfolio*

DLM Case Study – Archiving with Amazon

An image processing company IPC has provisioned their enterprise data lake on AWS. While the legacy data store had 10-year data, but data council has mandated data lake to hold 3-year worth of image and actions data. With respect to data archival, there could be two possible scenarios:

1. Data archival can be triggered via ageing policy

2. Data can be directly uploaded for archiving

In a typical DLM framework using AWS, the Amazon Glacier acts as the converging point in an archival strategy. In both of the above scenarios, data archives can be stored on AWS Glacier. In the first scenario, Amazon AWS allows lifecycle integration with S3 storage. Lifecycle rules can be defined and configured in AWS management console to move object data from S3 to Infrequent-access storage layer, and then to Glacier. Figure 5-8 shows the storage tiering within data lifecycle management starting from data creation in S3 to data archival in Glacier based on rules. All the lifecycle rules are checked as per their trigger settings. If the condition is met, data is progressively moved from one layer to another.

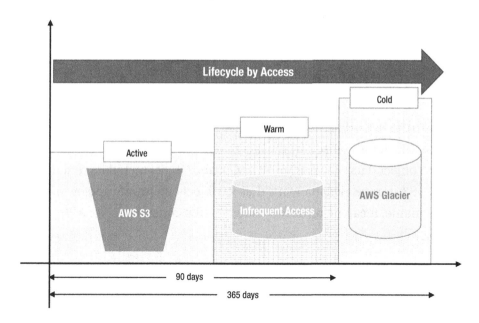

Figure 5-8. *Storage class tier support in data lifecycle management*

For the second scenario, legacy data can be directly uploaded into AWS Glacier through a direct upload technique. Predominantly, there are three methods to bulk upload the data into Glacier:

1. Data transfer over public internet via a secure tunnel

2. AWS Direct Connect for dedicated network bandwidth between site and AWS

3. AWS Snowball edge – AWS Snowball-edge is a high scale device with local compute and storage which is used for physical transfer of data from or into AWS. Equipped with an Amazon EC2 (m4.4xlarge) equivalent compute and storage, snowball can be used for cloud migration, disaster recovery, data center decommissions, and data proliferation.

Should DLM be always practiced through lifecycle rules? Can this process be exercised manually? Well, the benefit of S3 lifecycle managed Amazon Glacier is realized by the fact that index entries stay on S3, while only the object data moves from S3 to Glacier. Object metadata remains with the S3 storage, which means that an object can be referred by its user defined name. It can be retrieved using S3 APIs and not Glacier APIs.

Conclusion

Business are facing continuous pressure to store large amounts of data. Managing this data growth can be a real challenge for enterprises. There are multiple commercial archival solutions available in the market who assure of full-proof archival for an organization but what differentiates a successful and flop data archival is the data understanding. Archiving products may offer some drastic outputs in no time but issues due to a broken data archival strategy may pop up in the context of the future.

Key to data archival framework lies in the study of data dependencies, prioritization, and awareness with business SLAs. An efficient data archival strategy helps in optimizing the IT investments, improves data access performance, and increase returns on your investments.

In the next chapter, we are going to shift our focus to yet another critical component of data governance, namely, security. We will focus on data lake security principles and architect's considerations.

CHAPTER 6

Data Security in Data Lakes

"Data matures like wine, applications like fish."

—James Governor, Principal Analyst and founder of RedMonk

With enormous volume and high value of data, comes the responsibility to secure data from external intrusions and mitigate the chances of unwanted attacks. Every year, the world sees through ample cases of cyber thefts, security breaches, and digital attacks. As per Gartner's report[1] in Q1 2017, worldwide expenditure on security in 2017 was estimated to be $90 billion, which was 7.6% more than 2016 numbers. The need to have a robust security framework was well summarized by one of the Forrester researchers, who explained[2]:

> *Perimeter-based approaches to security have become out-dated. Security and privacy pros must take a datacentric approach to make certain that security travels with the data itself—not only to protect it from cybercriminals but also to ensure that privacy policies remain in effect.*

---

[1]Gartner press release, March 14, 2017 - https://www.gartner.com/newsroom/id/3638017

[2]https://www.forbes.com/sites/gilpress/2017/10/17/top-10-hot-data-security-and-privacy-technologies/

© Saurabh Gupta, Venkata Giri 2018
S. Gupta and V. Giri, *Practical Enterprise Data Lake Insights*,
https://doi.org/10.1007/978-1-4842-3522-5_6

In the last couple of years, security has strengthened its footprint in organizational information strategy. Top market players in data lines have nodded to the fact that a threat diagnosis holds a stronger relevance than threat mitigation. To prevent vulnerable areas from being exploited, organizations employ techniques like data encryption and redaction policies, proactive monitoring, and fine-grain access control. Figure 6-1 shows the key factors that play a vital role in determining a stable security strategy for an enterprise.

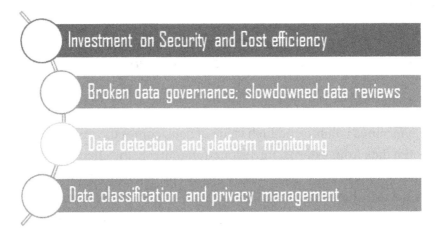

Figure 6-1. *Key factors that play a vital role in determining a stable security strategy*

This chapter takes a deep dive on to security aspects of a Hadoop data lake. It starts with a standard system architecture and familiarizes readers with components of network security, firewall, and Hadoop cluster.

System Architecture

In our earlier chapters, we have learned that Hadoop offers distributed storage and distributed processing paradigm to handle large data sets. While handling large clusters, the areas that we must keep in mind are the operating environment, its mode of operation, and network bandwidth. There are multiple factors that contribute to the choice of operating environment for Hadoop. Because of rapid advancement in the network and server technologies, Hadoop can live in few different environments.

1. Inhouse – Hadoop can be setup in the data center under the ownership and control of the business. They are setup with set of physical boxes (bare metal).

2. Managed – Managed setup is a variation of *inhouse* that consists of physical servers but not owned or operated by the organization. They are leased from an external vendor, who handles the provisioning, operations, and maintenance.

3. Cloud service models – This happens to be the talk of the town. Several technology giants have started offering virtualized servers for public use as "cloud" services. The virtualized infrastructure can be leveraged to run systems, as large as data center or as small as a simple application. As a cloud customer, companies pay what they subscribe and use; just like you enjoy buffet or al-a-carte lunch at restaurants.

Network Security

Network segmentation separates a cluster environment physically and logically, from a larger network. Physical network segmentation is separating the physical devices such as network switches, routers, VLANs, and firewalls, while the logical network segmentation could be achieved by using the separate Internet Protocol addresses.

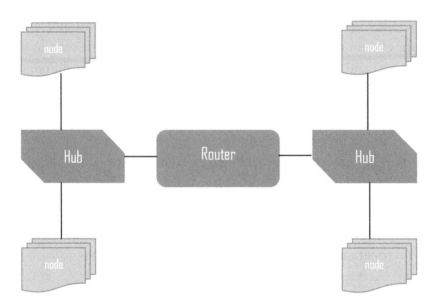

Figure 6-2. *Physical Segmentation with Router*

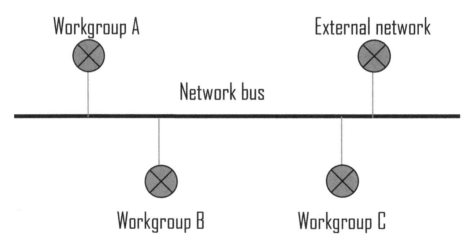

Figure 6-3. *Logical Separation with IP Subnets*

Network firewalls protect the internal network from unauthorized and malicious access from external agents. It can be configured to limit access to the outside from internal users. Firewall constantly monitors all incoming and outgoing traffic. It might restrict specific applications from accessing the network, block URLs from loading, and prevent traffic through certain network ports.

There could be network firewalls which block everything except the one (or few) which has been explicitly enabled for access. This practice helps in safeguarding the network from malicious threats. Firewall analyzes and hence protects the intranet based on the incoming IP, ports, and protocols.

Intrusion Detection and prevention is the process of monitoring the events occurring in your network and analyzing them for signs of possible incidents, violations, or imminent threats to your security policies. On the other hand, prevention is the process of detecting an intrusion and stopping such incident. These security measures are available as intrusion detection systems (IDS) and intrusion prevention systems (IPS), which become part of your network to detect and stop potential incidents. IDS logs can be captured, analyzed, and sent to monitoring systems for alerts. Intrusion detection can be signature based or anomaly based. Intrusion Prevention Systems monitors and tries to stop the intrusion.

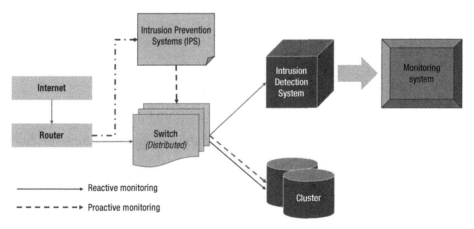

Figure 6-4. *Reactive monitoring versus Proactive monitoring using IPS and IDS processes*

Hadoop Roles within a cluster

Security policies can be applied to node groups in the cluster based on the roles each group plays and the services they provide. In a typical Hadoop Cluster, the various groups can be Master nodes, Worker nodes, Management Nodes, and Edge Nodes.

Master Node Group

This group includes all those nodes which host the master services like, Namenode, Standby name node, Resource Manager, Hive Server, and zookeeper. These services should be able to access other services, also since they are the important services, the security on these nodes should be high.

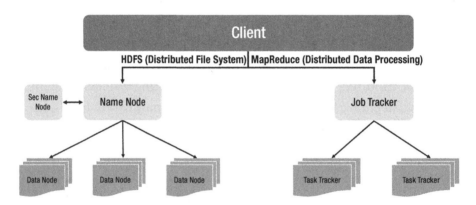

Figure 6-5. *Hadoop storage and processing framework*

A normal user/program should not be permitted to access these nodes, as it may cause the accidental vulnerabilities and situations such as confidential data access and service outages.

Worker Node Group

This group includes all those nodes which host the worker services such as Data Node services, Node Managers, and Task Trackers. Worker nodes need network level access to the respective master nodes as they communicate through RPC and TCP IP. Like master nodes, worker nodes also need a high security. Only the administrators should be allowed to access these nodes.

Management Nodes

Management nodes facilitate the seamless operation of the cluster by providing the configuration management, monitoring, and alerting. These nodes will typically contain the initial repositories from which the whole cluster will be built. It could be hive metastore, oozie, or Ambari server.

Apache Ambari is a management software of Hortonworks distribution. The server on which the Ambari is installed is a management node. It needs a backend database which again will be a management node. Since the security vulnerabilities on these nodes could cause the disturbance in the whole cluster, careful access management is needed.

Edge Nodes

The edge nodes host various clients with which the user interacts, such as Hive/Spark Clients, Gateways, Sqoop, Oozie clients, and Hue/Ambari views. Users can be classified based on their cluster usage.

Data security layers

For large enterprise systems, it is difficult to implement a data security solution that spans across different layers of data lake ecosystem. A system may offer multiple entry points like storage, network, and the user community. Therefore, it is a good practice to follow layered approach to secure all such points that can potentially expose vulnerability.

231

Let us go through the phases of data movement and see how unsecure points can be prevented from being attacked.

1. Data enters into lake via network – Network security layer prevents malicious attacks by authenticating users before they are allowed to access Hadoop data lake.

2. Storage – Data that resides within the lake is of huge business value and in order to secure this data, a strong encryption layer is a recommendable solution. We may go with Transparent Data Encryption (TDE), Ranger KMS, and many others who assure storage level encryption.

3. Data currently being accessed – The data which is selected from the lake should travel securely within data lake. Ranger offers wide variety of security features at the granular data level and can be used to implement dynamic column masking or row filtering for hive.

Host Firewalls for operating system security

Host firewalls can be iptables also called as Packet Filters, in Linux kernels, or the set of programs that intercept network traffic. A proxy server enables the connection on behalf of a specific network application from one network to another. Proxies are usually slower than packet filters.

IPTables policies can be configured to intercept the Hadoop cluster traffic. A typical Hadoop cluster will have certain common ports for different services. Iptables rules can be configured to restrict the traffic, through the IPs and these ports. The rules can be tightened or relaxed based on the requirement. For example,

```
iptables -A hdfs -p tcp -s 0.0.0.0/0 --dport 50070- -j ACCEPT
```

The above command will allow all the servers (0.0.0.0) to connect to the node, as the incoming IPs are not restricted. But the below one is stricter where it allows only a server with the IP 10.230.224.18 to connect.

```
iptables -A hdfs -p tcp -s 10.230.224.18/32 --dport 50070- -j
ACCEPT
```

Above command will be useful in a setup where all the communications to the cluster are done via edge node. The edge node IP must be added to the iptables policy.

Data in Motion

Data in motion is holds primary relevance due to its ability to drive actionable intelligence. However, beware of the fact that data is most vulnerable when in motion during the communication. It is the time during which data leaves the network and routes through unknown network components leaving it open for various forms of attack, hence data in motion must be protected.

Research has been published to tackle this problem and the core idea revolves around creating an encrypted channel. There are various layers of OSI stack on which this channel can be created but the most popular ones are built around Transport Layer of Open Systems interconnection (OSI) stack. Encrypting transport layer has benefits around encryption reuse across the applications.

Communication Problem

A typical communication can be showcased using scenario depicted by Figure 6-6.

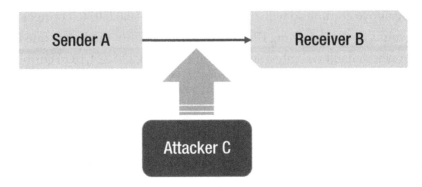

Figure 6-6. *Security attack on the communication channel set between A and B*

Sender A wants to send data to Receiver B securely. We don't want Attacker C to decipher or manipulate the communication between A and B. This can be achieved by encrypting the data between A and B so that C cannot understand and hence can't manipulate the data. Along with the confidentiality, it is also necessary that both A and B authenticate each other first before communicating as attacker C posing as either A or B can compromise the system.

Hence, there are 3 major problems which need to be addressed for a secured communication.

1. Authentication - Proving that you are who you claim to be

2. Confidentiality – Ensuring that only authenticated and intended user can read the data

3. Integrity – Ensuring that data is not tampered

How to solve this? Well, all three problems are currently being ensured by a class of algorithms called as Asymmetric key algorithms. It is also publicly known as public/private key cryptography.

A message encrypted with a public key can only be decrypted with a private key and vice versa. The private key is held as secret and the public key is open to public. This solves following problems

1. Authentication, which is done by Digital signatures. In this a message signed by sender using his private key. Anybody with sender's public key can decrypt the message and if successful can be confident that only the person with private key would have been able to encrypt that data.

2. Integrity is ensured by the encryption. The message becomes the input for the encryption algorithm and thus the decryption will only be successful if the intended message was same.

3. Confidentiality is maintained by signing the message with a person's public key. Hence the message can only be decrypted by a person who has the secret private key.

Two algorithms which are commonly used in the industry for this are RSA and Elliptic Curve Cryptography. Both are based on public / private key encryption. The keys by themselves are stored in files called as certificates. The most common of this are X.509 certificates which contain the public keys.

To digitally sign data hash functions are used. Hash functions have following properties.

When applied on a message the resulting message is very difficult to decipher and don't give a clue about the message hence impossible to reverse. Secondly, they have less collision factor, which is two messages with same hash function returns different results commonly used hash functions are MD5, SHA1, and SHA2.

The process goes like, sender creates message hash, encrypt it with private key. The receiver receives the message use hash function to generate the public key and decrypts the received message. If both the messages match than it can only be generated used the same hash functions also the one who possesses the private key would have only been able to encrypt the message hence authentication and integrity are maintained.

The validity of a public key is ensured by a centralized authority called as Certificate Authorities (CA). Public Key Infrastructure (PKI) maintains the validity and lifecycle of a certificate.

The problem with asymmetric cryptography is it is expensive for large data. The current infrastructure that is publicly available, can encrypt about 100kb of data efficiently, much lower than current market needs. To cater to this problem that are separate class of algorithms called as symmetric key algorithms where data is encrypted/ decrypted using the same key. They are much faster than asymmetric key algorithms. Common examples of these algorithms are AES (64- and 128-bit), Blowfish, DES (Internal Mechanics, Triple DES), Serpent, and Twofish. The problem with these method is to give same key to both the parties which is very difficult task between two unknown parties. It is formalized by exchange called as Diffie Hellman.

The overall idea is to use both the class of algorithms for their strong points. We generally use asymmetric key algorithms for authentication and passing the symmetric key to both the parties. And then encrypt bulk of the messages with the symmetric keys that were passes between each other.

This strategy is implemented at transport layer using protocols like TLS and IPSec. TLS is successor of the old methodologies called SSL. These strategies play an important role in giving security to data in motion.

Data at Rest

Data at Rest refers to the data lying physically in a data store, warehouse, archives, online or offline backups, or any other device. It may or may not be under active operation but serves a decent purpose of drawing historical insights. LUKS or Linux Unified Key Setup, can be used to encrypt data at rest on a disk. It specifies a platform-independent standard on-disk format which provides compatibility via standardization. The reference implementation for LUKS operates on Linux and is based on enhanced version of cryptsetup, using dm-crypt as the disk encryption backend. Dm-crypt is a transparent disk encryption subsystem which uses cryptographic routines from the Linux kernel's Crypto API. This includes most popularly used encryption algorithms such as AES, hash functions such as SHA-256.

We need to encrypt data resting on a disk to prevent physical/offline attacks on the devices. The disks may be remote locations, which makes it necessary for us to anticipate and prevent attacks such as theft of disks by external organizations. In an event where a LUKS-encrypted disk is stolen, an attacker will be unable to mount this device without the valid passphrase. Mounting the device through cryptsetup, will also require a passphrase. Due to the protection provided by LUKS in insisting on strong passwords, as well as use of salts to increase security against low-entropy passwords, most of the attacks faced by other key management systems are eliminated. LUKS also provides support for up to 8 passphrases for a disk (Figure 6-7).

Figure 6-7. *Disk layout of LUKS*

From the above image, KM stands for [Key Material], while bulk data is the user data.

Procedure to generate and verify key in LUKS

The user provides a passphrase. This passphrase is appended with a salt value. The password based key derivation function (PBKDF) generates a key from this value. With the help of encrypted master key, it is converted into the actual master key (MK).

For the verification of key, we provide master key and the salt values to PBKDF2 function and compare with the master key digest value as shown in Figure 6-8.

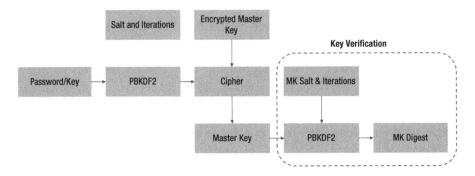

Figure 6-8. *Key generation and verification in LUKS*

Access flow for the user

The user interacts with the mapper to the physical disk through the mount point upon successful authentication through cryptsetup (Figure 6-9).

Figure 6-9. *User access flow through cryptsetup*

Below, we have used b-tree file system (btrfs) to measure read/write performances before and after LUKS encryption. The below steps give information on installation of btrfs, creating a mount point and checking read/write performances.

Step 1 - Install btrfs

```
$ yum install btrfs-progs
```

Step 2 - Create btrfs filesystem on disk

```
$ mkfs.btrfs /dev/sda1
```

Step 3 - For testing purpose, we mount this to /mnt

```
$ mount /dev/sda1 /mnt
$ chmod 777 /mnt
```

Step 4 - checking for write speed (file size: 948MB, total size allocated to file system was 1GB, so we could not try for larger files)

```
$ dd if=/dev/sdf/test1 of=/mnt/dd-test1 bs=1M conv=fsync
Result -  72MB/s
```

Running the same command again,

```
Result -  69MB/s
```

Step 5 - Checking for read speed(file size:948MB)

```
$ dd if=/mnt/dd-test1 of=/dev/null bs=1M
Result -  79.6MB/s
```

Running the same command again,

```
Result: 3.3GB/s
Reason: Data is cached.
```

For using LUKS, follow the below steps:

Step1 - Unmount the filesystem

```
$ umount /dev/sda1
```

Steps to encrypt disk using LUKS:

1. Install cryptsetup on machine

    ```
    $ yum install cryptsetup-luks
    ```

2. The above command will install cryptsetup – which
    is the frontend we will use to encrypt the disk with
    LUKS.

Setting up LUKS on the disk /dev/sda1:

1. Format the disk, take necessary backup beforehand.

    ```
    $ sudo cryptsetup luksFormat /dev/sda
    ```

2. The first time to setup luks encryption, we use
    the command luksOpen as follows. The output
    of this command is a prompt for the user to enter
    passphrase. This is the key which will be in Key
    slot 0.

    ```
    $ sudo cryptsetup luksOpen /dev/sda1 crypt
    ```

3. We then use the mkfs command with the specified
    filesystem – since we have btrfs(B-tree filesystem)
    on /dev/sda1, we use mkfs.btrfs to create the
    filesystem mapping as follows: (dev/mapping is the
    directory where we store the logical mapping of the
    filesystem)

    ```
    $ sudo mkfs.btrfs /dev/mapper/crypt
    ```

4. We now mount the filesystem from the mapper as
    follows:

    ```
    $ sudo mount /dev/mapper/crypt /mnt
    ```

5. We add necessary permissions to the directory using chmod:

```
$ sudo chmod 777 /mnt
```

After LUKS encryption:

1. After encrypting the disk /dev/sda1 with Luks, check for luks dump to see the key slots which are active and cipher text

```
$ cryptsetup luksDump /dev/sda1
```

2. Output of luksDump is seen in Figure 6-10.

```
Cipher name:     aes
Cipher mode:     xts-plain64
Hash spec:       sha256
Payload offset: 4096
MK bits:         256
MK digest:       34 88 f5 03 ce fc 30 61 85 3b b8 eb 9f cb 15 2e 58 80 c9 44
MK salt:         7b ff 61 d7 ce 99 d3 e3 92 56 f7 d7 e9 a6 8e a3
                 45 99 dc bb 38 e6 a6 6f b2 b0 09 f1 6f a2 bb e4
MK iterations:   75750
UUID:            c892b7a7-49dd-45de-94ac-2befc293fcf7

Key Slot 0: ENABLED
        Iterations:              596041
        Salt:                    53 ec ef 28 aa 2c 7d f8 d2 ae 8a 20 5e c4 1a 3f
                                 33 88 8e f4 a9 7a 07 f1 0f 7d 92 34 e3 98 39 70
        Key material offset:     8
        AF stripes:              4000
Key Slot 1: DISABLED
Key Slot 2: DISABLED
Key Slot 3: DISABLED
Key Slot 4: DISABLED
Key Slot 5: DISABLED
Key Slot 6: DISABLED
Key Slot 7: DISABLED
[root@localhost dev]# _
```

Figure 6-10. *LUKS dump output*

3. As sda1 is encrypted, mounting it is not possible

    $ mount /dev/sda1 /mnt gives us the error: "unknown filesystem type:crypto-LUKS"

4. We have to open /dev/sda1 with cryptsetup, it requires the passphrase provided earlier

    $ cryptsetup open /dev/sda1 crypt – this command requests a passphrase, providing the wrong passphrase makes it prompt again, Correct passphrase lets the device be accessible through the logical mapping that we have named crypt.

5. Mount /dev/sda1 through the logical volume /dev/ mappercrypt

    $ mount /dev/mapper/crypt /mnt

6. You can now access the files on sda1 through the mount point /mnt.

```
Key Slot 4: DISABLED
Key Slot 5: DISABLED
Key Slot 6: DISABLED
Key Slot 7: DISABLED
[root@localhost dev]# ls /mnt
[root@localhost dev]# mount /dev/sda1 /mnt
mount: unknown filesystem type 'crypto_LUKS'
[root@localhost dev]# cryptsetup open /dev/sda1 crypt
Enter passphrase for /dev/sda1:
No key available with this passphrase.
Enter passphrase for /dev/sda1:
[root@localhost dev]# mount /dev/mapper/crypt /mnt
[root@localhost dev]# ls /mnt
dd-abc  dd-test1
[root@localhost dev]# cat dd-abc
cat: dd-abc: No such directory
[root@localhost dev]# cat /mnt/dd-abc
bcdefghijklmnopqrstuv
wxyz12
345678
[root@localhost dev]# umount /mnt
[root@localhost dev]# cryptsetup close crypt
[root@localhost dev]# cat /mnt/dd-abc
cat: /mnt/dd-abc: No such file or directory
[root@localhost dev]# cat /mnt/dd-abc_
```

Figure 6-11. *Using cryptsetup to access the device*

Performance using LUKS

Below commands can give you read and write speed of LUKS implementation. Note that these numbers are system dependent and may show different values on your system.

Checking for write speed:

```
$ dd if=/dev/sdf/test1 of=/mnt/dd-test1 bs=1M conv=fsync
Result -  71MB/s
```

Checking read speed:

```
$ dd if=/mnt/dd-test1 of=/dev/null bs=1M
Result -  91MB/s
```

Multiple passphrases with LUKS

To add a new key, use the following command. It requires one of the existing passphrases to allow setting a new passphrase.

1. `$ cryptsetup luksAddKey /dev/sda1` prompts for existing passphrase and then allows us to setup a new one. This will make Key Slot 1 active. Run `luksDump` command again to check active key slots.

2. We can set up to 8 passphrases for a single device.[3]

---

[3]https://wiki.archlinux.org/index.php/Dmcrypt/Device_encryption#
Cryptsetup_passphrases_and_keys

Kerberos

Kerberos is a computer network and authentication protocol that works based on tickets to allows nodes to communicate over a non-secure network to prove their identity to one another in a secure manner. In simple terms, it employs strong authentication methods to establish a user's identity to allow secure access in data lake.

Users, devices, and services that deploy Kerberos security framework need only to trust the Key Distribution Centre (KDC). It runs as a single process and offers two services, namely, an authentication service and a ticket granting service. KDC tickets enable mutual authentication which allows nodes to securely prove their identity to one another.

Kerberos derives its name from a Greek mythology character Kerberos, the three-headed guard dog of Hades. Primarily, it was designed for client-server model to provide mutual authentication to users and servers for identity verification. Kerberos was originally developed for Project Athena at the Massachusetts Institute of Technology (MIT).

Kerberos Protocol overview

1. The identities on Kerberos is called principals. Every user and service that participates in the Kerberos authentication protocol requires a principal to uniquely identify itself.

2. Principals are of two categories.

    a. User principals - User principals relate to username and accounts in a OS.

    b. Service principals - Service principals represents services that a user need to access, such as a specific server or a database.

3. Kerberos realms - A Kerberos realm is an authentication administrative domain. All the principals are assigned to a realm.

4. Key Distribution center (KDC) - The KDC has three components.

    a. Kerberos database - It stores all the information about the principal and realm they belong to, along with other information.

        i. Kerberos principals in the database are stored with the below naming convention:

            1) greg@TEST.COM - A user principal that is distinctly identifies the user - Greg in the realm TEST.COM. Always the realm name is in the upper case.

            2) Julie/admin@TEST.COM - A different way to put the user principal. Here the administrator Julie in the realm TEST.COM. The "/" (slash) separates the short name and the admin.

            3) hdfs/node02.test.com@TEST.COM - This is a example of server principal from the HDFS service on the host node02.test.com in the TEST.COM.

    b. Authentication service - An authentication service, also known as a ticket-granting ticket (TGT), is a small amount of encrypted data that is issued by a server in the Kerberos authentication model to begin the authentication process. When a client receives an authentication ticket, it sends back the ticket to the KDC server along with its identity information. The KDC server generates a service ticket and a session key (which includes a form of password), thereby completing the authorization process for that session.

c.   Ticket granting service (TGS) - TGS is responsible for validating ticket-granting tickets and granting service tickets. Service tickets enable authenticated principals to use the service provided by the application server and identified by the server principal.

Kerberos components

Let us check out the Kerberos components before discussing the flow.

1.   TGT: Ticket Granting Ticket or Ticket to Get Tickets (TGT) is a small, encrypted identification file with a limited validity period. After authentication, this file is granted to a user to establish a secure client-server session for the needed service.

2.   Kerberos Principal - A principal is an identity in the cluster for any service, node or user.

3.   Kerberos Realm - The term realm indicates an authentication administrative domain, that defines a group of systems that are under the same master KDC (Key Distribution Center/Kerberos Domain Controller) Server. KDC server runs two functional service.

a.   AS - Authentication Service – Authenticates Kerberos principals

b.   TGS: Ticket Granting Service - Grants access to specific services

Kerberos flow

A user Greg is trying to connect to a domain test.com on a particular node (node02) through a service. Let us understand how Kerberos will authenticate Greg before allowing him to access the desired domain.

Let us distinctly identify the kerberos components from our example.

- Kerberos Realm is the domain, i.e., TEST.COM.

- USER of a System with Kerberos user principal greg@TEST.COM, i.e., Greg

- A service within the cluster that will be hosted on node, node02.test.com identified by testservice/node02.test.com@TEST.COM

The KDC Server for the Kerberos realm TEST.COM - kdc.test.com

1. Greg needs to obtain a TGT. To do this, he initiates a request to the AS at kdc.test.com, identifying himself as the principal greg@TEST.COM.

2. The Authentication service responds by providing a TGT that is encrypted with Greg's secret key.

3. Upon receipt of the encrypted message, Greg is prompted to enter the correct password for the principal greg@TEST.COM in order to decrypt the message.

4. After successfully decrypting the message containing the TGT, Greg now requests a service granting ticket from the TGS at kdc.test.com for the service within the realm (TEST.COM) identified as hdfs/node02.test.com@TEST.COM, presenting the TGT along with the request.

5. The TGS validates the TGT and provides Greg a service ticket encrypted with the `hdfs/node02.test.com@TEST.COM` principal key

6. Greg now presents the service ticket to server hosting the service, which then decrypt it using the `hdfs/node02.test.com@TEST.COM` key and validate the ticket. Thus, the server allows access to the client (Greg) to use the service by establishing a client-server session after a successful authentication has been achieved.

Kerberos Realm for bigger organizations - While a single realm works well for an organization but often but it often not realistic for some bigger enterprises. Over a period, larger organizations end up setting multiple realms just to simplify and to segregate different part of the organization. By default, KDC is known for its own realm, principals and database. If a user from one realm wants to use a service that is controlled by another realm then a Kerberos trust is needed between the two realms. Let's say a FINANCE and an HR professional need to talk to each other then they need to trust the information of each other realms.

So basically, there are two kinds of trust, one-way and two-way trust. Let's say Finance realm needs to access the HR realm. This scenario requires one-way trust. To establish two-trust, i.e., full trust, the principals need to exist on both the realms. For example - for the HR.TEST.COM realm to have a full trust with the `FINANCE.TEST.COM` realm, both the principals `krbtgt/FINANCE.TEST.COM@HR.TEST.COM` and `kbrtgt/HR.TEST.COM@FINANCE.TEST.COM`.

Kerberos commands

Below is the list of Kerberos administrative commands.

1. KINIT – kinit obtains and caches an initial ticket-granting ticket for principal.

2. Kinit using keytab file – A keytab is a file containing pairs of Kerberos principals and encrypted keys (which are derived from the Kerberos password). You can use a keytab file to authenticate to various remote systems using Kerberos without entering a password. However, when you change your Kerberos password, you will need to recreate all your keytabs. Keytab files are commonly used to allow scripts to automatically authenticate using Kerberos, without requiring human interaction or access to password stored in a plain-text file. The script is then able to use the acquired credentials to access files stored on a remote system.

3. KLIST - klist lists the Kerberos principal and Kerberos tickets held in a credentials cache, or the keys held in a keytab file.

4. KDESTROY - The kdestroy utility destroys the user's active Kerberos authorization tickets by overwriting and deleting the credentials cache that contains them. If the credentials cache is not specified, the default credentials cache is destroyed.

Kerberos principles mapping to usernames – Kerberos uses two principals (greg@TEST.COM) or three parts (hdfs/node02.test.com@TEST.COM), that contains short name, realm and optional instance name or hostname. To simplify working with usernames, Hadoop maps Kerberos principal names to local usernames by using auth_to_local setting in the krb5.conf file, or Hadoop specific rules can be configured in the Hadoop.security. auth_to_local parameter in the core-site.xml.

A mapping consists of a set of rules that are evaluated in the order listed in the Hadoop.security.auth_to_local property. The first rule that matches a principal name is used to map that principal name to a short name. Any later rules in the list that match the same principal name are ignored. You specify the mapping rules on separate lines the Hadoop. security.auth_to_local property as follows:

```
<property>
  <name>Hadoop.security.auth_to_local</name>
  <value>
  RULE:[<principal translation>](<acceptance filter>)
       <short name substitution>
  RULE:[<principal translation>](<acceptance filter>)
       <short name substitution>
  DEFAULT
  </value>
</property>
```

Hadoop user to group mapping – The groups of a user are determined by a group mapping service provider. Hadoop supports various group mapping mechanisms, configured by the Hadoop.security.group. mapping property. This means that only the groups that are configured on the server where the mapping is called are visible to Hadoop. In practice it is very important for all the servers in your Hadoop cluster to have a consistent view of the users and groups that will be accessing the cluster.

```
<property>
<name>Hadoop.security.group.mapping</name>
<value>org.apache.Hadoop.security.LdapGroupsMapping</value>
</property>
```

Users to group mapping using LDAP - For the environment where the groups are only available from the LDAP or Active Directory server not from the cluster nodes. Hadoop provides *LdapGroupsMapping* implementation. This method can be configured by setting parameters in the core-site.xml on the namenode, jobtracker, or resourceManager.

This provider supports LDAP with simple password authentication using JNDI API. The parameter Hadoop.security.group.mapping.ldap. url must be set. This refers to the URL of the LDAP server for resolving user groups.

The Hadoop.security.group.mapping.ldap.base configures the search base for the LDAP connection. This is a distinguished name, and will typically be the root of the LDAP directory. Get groups for a given username first looks up the user and then looks up the groups for the user result. If the directory setup has different user and group search bases, use the parameters Hadoop.security.group.mapping.ldap.userbase and Hadoop.security.group.mapping.ldap.groupbase configs.

If the LDAP server does not support anonymous binds, set the distinguished name of the user to bind in Hadoop.security.group. mapping.ldap.bind.user parameter. The path to the file containing the bind user's password is specified in Hadoop.security.group.mapping. ldap.bind.password.file. This file should be readable only by the Unix user running the daemons.

It is possible to set a maximum time limit when searching and awaiting a result. Set Hadoop.security.group.mapping.ldap.directory.search. timeout to 0 if infinite wait period is desired. Default is 10,000 milliseconds (10 seconds). This is the limit for each LDAP query. If Hadoop.security. group.mapping.ldap.search.group.hierarchy.levels is set to a positive value, then the total latency will be bounded by max(Recur Depth in LDAP, Hadoop.security.group.mapping.ldap.search.group.hierarchy. levels)*Hadoop.security.group.mapping.ldap.directory.search. timeout.

The Hadoop.security.group.mapping.ldap.base configures how far to walk up the groups hierarchy when resolving groups. By default, with a limit of 0, in order to be considered a member of a group, the user must be an explicit member in LDAP. Otherwise, it will traverse the group hierarchy Hadoop.security.group.mapping.ldap.search.group.hierarchy. levels levels up.

```
<property>
<name>Hadoop.security.group.mapping.provider.ad4usersX.ldap.
url</name>
<value>ldap://ad-host-for-users-X:389</value>
  <description>
    ldap url for the provider named by 'ad4usersX'. Note this
    property comes from
    'Hadoop.security.group.mapping.ldap.url'.
  </description>
</property>

<property>
<name>Hadoop.security.group.mapping.provider.ad4usersY.ldap.
url</name>
<value>ldap://ad-host-for-users-Y:389</value>
  <description>
    ldap url for the provider named by 'ad4usersY'. Note this
    property comes from
    'Hadoop.security.group.mapping.ldap.url'.
  </description>
</property>
```

Hadoop Users – In a Hadoop environment all the Hadoop users of a cluster must be provisioned on all the servers of the cluster. These users can exist on the local /etc/passwd password file or, more commonly can be provisioned by having the servers access a network based directory service like open-LDAP or Active Directory.

Authentication – If Hadoop is configured with all its defaults, Hadoop doesn't do any authentication of users. This is an important realization to make, because it can have serious implications in a corporate data centre.

Let's say Greg User has access to a Hadoop cluster. So far, no security regulations have been imposed on the Hadoop cluster. Users can interact without any authentication. Although Greg is neither a superuser nor has hdfs user password, but he has access to the client machine which is configured to access the cluster. Irresponsibly, he issues two commands:

```
sudo useradd hdfs
sudo -u hdfs Hadoop fs -rmr /
```

Needless to say, the cluster has gone off and deleted everything. So, what has just happened? In an unsecured cluster, by default, NameNode or JobTracker don't require any sort of authentication. This implies that you can do all those operations that fall under the bucket of hdfs and mapred users.

In a distributed system, it is important that all requests by a user is validated by user identity. We need to authenticate every interaction. For example, in a mapreduce job, the authentication happens between the client and the namenode and between client and the job tracker. In order to submit the job, a jobtracker then creates multiple tasks that are launched by each taskTracker in the cluster. Each tasktracker has to communicate with the namenode in order to open the files that make up its input split. For the NameNode to enforce filesystem permissions, each task should authenticate against the NameNode. Hadoop adopts token-based authentication approach to whitelist a client and allow it to issue an action on the cluster.

Hadoop solves this problem by issuing authentication tokens that can be distributed to each task but are limited to a specific service. Let us check how this delegation tokens work –

1. A client issues an RPC to request a delegation token via Kerberos ticket for authentication.

2. NameNode receives and responds with a delegation token.

3. Once authentication is done, client is allowed to issue an action using the delegation token for authentication.

4. After the token gets validated, NameNode acts to the command issued by the client

Authorization – Authorization is an approach to define what you can access and what not. Remember authentication gives a technique to prove one's identity, while authorization is a post-authentication activity that justifies your access rights within a Hadoop cluster. In HDFS authorization is realized through file permissions.

If you run ls -l in a directory, you will get the listing as below.

```
[etl@ip-etl]$ Hadoop fs -ls /apps/hive/warehouse/db/
Found 3 items
drwxrwxrwx    -etl    hdfs    0 2018-02-22 10:13 /apps/hive/
warehouse/db/employees
drwxrwxrwx    -etl    hdfs    0 2018-02-22 10:13 /apps/hive/
warehouse/db/departments
drwxrwxrwx    -etl    hdfs    0 2018-02-22 10:13 /apps/hive/
warehouse/db/locations

[etl@ip-etl]$ Hadoop fs -ls /apps/hive/warehouse/db/employees
Found 1 items
-rwxrwxrwx    3 hive hdfs 21839221 2018-02-25 13:40 /apps/hive/
warehouse/db/employees/000000_0
```

How a client authorizes its access to a block situated on a datanode? It is done through a block access token mechanism. A standard authorization process goes through the following steps:

1. An *authenticated* client issues a read request

2. NameNode gathers the information about the data blocks, data nodes, and the closest data node from the meta-information memory structure

3. NameNode sends the block address information to the client along with the block access token

4. Client selects the closest data node based on the information received from the NameNode

5. Client requests the block with block access token from the closest data node

6. Data node that receives the request verifies authenticated information of the requested block

7. Token authenticator is created out of block access token and secret key shared by the NameNode. Secret key is an authentication token between a NameNode and data node. It is renewed regularly and shared by NameNode via heartbeat communication channel. The secret key encrypts the block access token requested by a client.

8. Created and Received token authenticators are compared. When matched, the requested block is sent to the client.

The Hadoop Distributed File System (HDFS) implements a permissions model for files and directories that shares much of the POSIX model. Each file and directory is associated with an owner and a group.

The file or directory has separate permissions for the user that is the owner, for other users that are members of the group, and for all other users. For files, the r permission is required to read the file, and the w permission is required to write or append to the file. For directories, the r permission is required to list the contents of the directory, the w permission is required to create or delete files or directories, and the x permission is required to access a child of the directory.

In contrast to the POSIX model, there are no *setuid* or *setgid* bits for files as there is no notion of executable files. For directories, there are no setuid or setgid bits directory as a simplification. The sticky bit can be set on directories, preventing anyone except the superuser, directory owner or file owner from deleting or moving the files within the directory. Setting the sticky bit for a file has no effect. Collectively, the permissions of a file or directory are its mode. In general, Unix customs for representing and displaying modes will be used, including the use of octal numbers in this description. When a file or directory is created, its owner is the user identity of the client process, and its group is the group of the parent directory

HDFS ACL

With the release of Hadoop 2.4, we can now use extended ACL's. These ACL's work very much the same way as in any unix OS. ACLs are useful for implementing permission requirements that differ from the natural organizational hierarchy of users and groups. An ACL provides a way to set different permissions for specific named users or named groups, not only the file's owner and the file's group.

By default, support for ACLs is disabled, and the NameNode disallows creation of ACLs. To enable support for ACLs, set `dfs.namenode.acls. enabled` to true in the NameNode configuration.

HDFS Authorization with Apache Ranger

Apache Ranger provides a user synchronization utility to pull users and groups from Unix or from LDAP or Active Directory. The user or group information is stored within Ranger portal and used for policy definition.

HDFS is core part of any Hadoop deployment and to ensure that data is protected in Hadoop platform, security needs to be baked into the HDFS layer. HDFS is protected using Kerberos authentication, and authorization using POSIX style permissions/HDFS ACLs or using Apache Ranger.

Apache Ranger is a centralized security administration solution for Hadoop that enables administrators to create and enforce security policies for HDFS and other Hadoop platform components. Apache Ranger offers a federated authorization model for HDFS. Ranger plugin for HDFS checks for Ranger policies and if a policy exists, access is granted to user. If a policy doesn't exist in Ranger, then Ranger would default to native permissions model in HDFS (POSIX or HDFS ACL). This federated model is applicable for HDFS and Yarn service in Ranger.

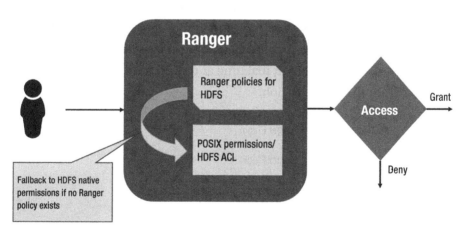

Figure 6-12. *Apache Ranger's federated authorization model*

For other services such as Hive or HBase, Ranger operates as the sole authorizer which means only Ranger policies are in effect. The option for fall back model is configured using a property in Ambari ➤ Ranger ➤ HDFS config ➤ Advanced ranger-hdfs-security.

What Ranger does?

Apache Ranger offers a centralized security framework to manage fine-grained access control across HDFS, hive, hbase, storm, Knox, Solr, Kafka, Nifi, and Yarn. Apache Ranger console, security administrators can easily manage policies for access to files, folders, databases, tables, or column. These policies can be set for individual users or groups and then enforced consistently across Hadoop stack.

The Ranger Key Management Service (Ranger KMS) provides a scalable cryptographic key management service for HDFS "data at rest" encryption. Ranger KMS is based on the Hadoop KMS originally developed by the Apache community and extends the native Hadoop KMS functionality by allowing system administrators to store keys in a secure database.

Ranger also provides security administrators with deep visibility into their Hadoop environment through a centralized audit location that tracks all the access requests in real time and support multiple destination sources including HDFS and Solr.

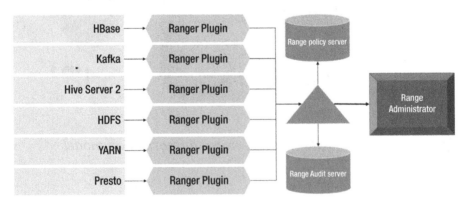

Figure 6-13. *Ranger setup for Hadoop stack*

Ranger Admin – The ranger admin portal provides an interactive interface for security administration. Security admins can create or modify policies via portal. It has an audit server that collects and ships audit data from plugins to HDFS database.

Ranger Plugin – Ranger plugins are java-based programs which are embedded within each component process. Figure 6-13 shows individual ranger plugin embedded within hive server2, YARN, Kafka, or Presto. Plugins pull policies from a central over and store it locally. During the event of a user request, plugin intercept the request for evaluation against the policy definition. It is also responsible for sending data to the audit server.

User group sync – Ranger helps with the user synchronization by pulling users and groups from LDAP or active directory and storing them within ranger.

Conclusion

The chapter talks about data lake security in an offbeat manner. The reason for being offbeat is that it doesn't dive into the layers of data masking in hive or password protection. Instead, it makes a fair attempt at developing the security awareness and thought process behind architecting a security framework for data lakes. We discussed the security aspect of data in motion as well as at rest. The use-case showcasing Kerberos security framework will help the readers to understand how Kerberos helps in establishing identity for clients, hosts, and services, without any chance of network leakage. Not just that, it can be extended and integrated with other identity management tools like LDAP and active directory.

In the next chapter, we are going to cover high availability of data lakes. Not just the high availability to Hadoop components, we'll also see how to setup multi-site data sites in active-passive and active-passive configurations.

CHAPTER 7

Ensure High Availability of Data Lake

"When human judgment and big data intersect, there are some funny things that happen"

—Nate Silver, founder and editor in chief of FiveThirtyEight

Such is the power of data analytics, that enterprises are almost resting on to the daily nuggets of information that can unlock and drive new business opportunities. The art and exercise of data accumulation, real-time processing, and data crunching help businesses with the most distilled format of information. It keeps them at pace with the market information, understand industry trend and act fast. With such a dependency on day to day life with data, organizations pay utmost attention towards support functions of enterprise data lake. Data lake support functions include data quality, governance, architecture, and administration. One of the administrative aspects of data lake is availability and disaster recovery.

© Saurabh Gupta, Venkata Giri 2018
S. Gupta and V. Giri, *Practical Enterprise Data Lake Insights*,
https://doi.org/10.1007/978-1-4842-3522-5_7

Why availability of data lake is critical to the business? Think of a heavy part production unit whose assembly line is dependent on data which is an analytical function of enterprise resource planning, data warehouse, customer services, and few statistical models. If the consumption layer data is stale or unavailable, it may incur direct impact on the business. Similarly, if retail analytics run deep, the learning model on consumer behavior and releases purchase vouchers to potential buyers, it will fail to contemplate if data lake is unavailable. In another peculiar case of data-driven forecasting, regular and in-advanced seismic updates are required by weather forecasters to act swiftly, ensure safety, minimize capital losses, and plan remedy actions well in advance. Data lake must be available to certain the continuity of daily analytical insights.

Disaster recovery is yet another term that gets associated with high availability of a data lake ecosystem. As "data lakes" become more business-critical in nature, grow in terms of volume, multi-fold data ingestion and egression, and operate in continuous streams, conventional disaster recovery strategies no longer go well with enterprises. The fact is imperative that organizations are implementing a data lake disaster recovery plan that can prepare for outages during failure events.

This chapter will shift gears to one of the crucial aspect of data governance, i.e., data availability. We will discuss high availability patterns of a Hadoop cluster and what could be the strategy to mitigate risks during outages. We will cover disaster recovery strategies in the context of data lake.

Scale Hadoop through HDFS federation

Data scaling is one thing which is gripping within the organizations who are dealing with deluge of data sets. A scalable data lake builds strong immunity against the meteorically growing and rapidly changing data sets. If you hear your users complaining about service outages or slow performance due to high disk usage or low memory, you must be ready with the toolkit for data scaling. We assume that the readers would be aware of scale-up and scale-out approaches of data scaling.

Before getting into the intricacies of Hadoop high availability, let us understand how Hadoop maintains and promotes the quotient of scalability in business-critical environments. Hadoop 1 was, by far, presented a suboptimal framework that assured of restrained availability and scalability. Some of the major challenges with Hadoop 1 architecture are listed below.

1. Availability constraints due to single NameNode – Single NameNode exposes the risk of single point of failure. If it fails, Hadoop cluster goes through an outage until NameNode is brought up.

2. Scalability – Hadoop DataNodes are well scalable and data can cut across multiple data nodes depending upon block size and replication factor. However, it can have a single namespace. A NameNode holds the namespace volume in memory, which is opaque to data nodes. To accommodate a larger namespace volume, NameNode can only be scaled vertically. Impact of limited scalability are listed as below.

    a. Ability to contain namespace volume of large size and long duration is restricted by the compute capacity of NameNode.

    b. All data access requests are served by single NameNode. NameNode and hence system's performance used to depend upon its throughput capacity.

A NameNode acts as the brain of Hadoop file system. It manages a *namespace volume* which facilitates an abstraction layer over namespace and physical storage layer. Key points include:

- Namespace holds the metadata of Hadoop file system, directories, and blocks.

- Block pool is the block management layer through which a NameNode carries out critical block operations like create, update, delete, underreplication, overreplication, and block reporting.

  - Block pool neither store data blocks nor data blocks; they are physically stored in data nodes. Data nodes send heartbeats to the NameNode on periodic basis.

Hadoop 2 addressed above challenges by introducing two changes. First, there is a provision for a standby NameNode to survive failures in active NameNode. The editLogs maintained by active NameNode are replicated to standby NameNode and in case of outage, standby NameNode assume the role of primary NameNode.

Second, Hadoop 2 introduced a federated architecture of multiple NameNodes to share namespace volumes among them. Each NameNode manages a chunk of namespace volume and is isolated from other contemporary NameNodes. HDFS or NameNode federation brings the ability to scale horizontally and mitigates the memory constraint at NameNode level. Namespace volume comprises of a namespace and pool of blocks pertaining to that namespace.

Figure 7-1 shows the NameNode federation setup.

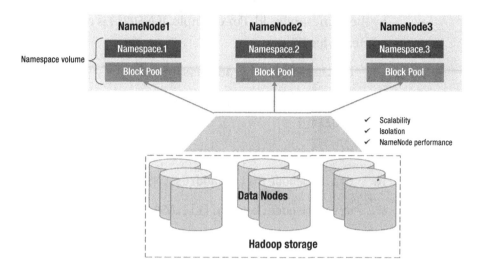

Figure 7-1. *HDFS federation architecture*

Key highlights and design considerations:

1. NameNode federation doesn't hinders the working architecture of Hadoop. Each active NameNode has its standby counterpart.

2. If active NameNode goes down, standby NameNode takes over and manages respective namespace volume

3. If both active and standby NameNode undergo a failure, only the pool of blocks managed by a NameNode become inaccessible

4. Configuration

   a. Specify comma separated NameNode in `dfs.nameservices` parameter. This should be done in the event of initial configuration or addition of a NameNode to the cluster.

265

b. Configuration of NameNode, secondary
    NameNode, and backup node must be suffixed
    by the nameservice id. For example, if *NS0214* is a
    nameservice ID,

    i.   `dfs.namenode.rpc-address.NS0214`

    ii.   `dfs.namenode.servicerpc-address.`
        `NS0214`

    iii.   `dfs.namenode.http-address.NS0214`

    iv.   `dfs.namenode.https-address.NS0214`

    v.   `dfs.namenode.keytab.file.NS0214`

    vi.   `dfs.namenode.name.dir.NS0214`

    vii.   `dfs.namenode.edits.dir.NS0214`

    viii.   `dfs.namenode.checkpoint.dir.NS0214`

    ix.   `dfs.namenode.checkpoint.edits.dir.`
        `NS0214`

c. When adding a new NameNode to the cluster

    i.   Add nameservice parameter and modify
        configuration file

    ii.   Start primary and secondary namenodes

    iii.   Refresh datanodes to identify new
        NameNode

```
$HADOOP_PREFIX/bin/hdfs dfsadmin -refreshNameNodes
<DN_host>:<DN_rpcPort>
```

High availability of Hadoop components

In our earlier chapter, we have discussed the Hadoop storage as well as processing architecture. From the storage perspective, NameNode becomes the gatekeeper of all read and put requests. Similarly, from a processing standpoint, Hadoop offers multiple entry points to enable data processing as per the skill knowhow, expertise with data-play, and ability to align data exercises with service level agreements. This section will highlight high availability configuration for different components of Hadoop.

Hive metastore

Hive metastore is the repository that contains metadata of all the tables created in hive. It is used to put data in shape (or schema) while reading data from the Hadoop cluster. Other processing frameworks like spark, Cloudera Impala, and Oracle Big Data SQL can also leverage hive metastore for schema readiness. Therefore, it becomes critical to ensure that the metastore service is available to its consumers.

Standby metastore – Hive can be configured on hosts, where metastore replication should happen. All the hosts can be specified as a list in a configuration property `hive.metastore.uris` in /etc/hive/conf.server/ hive-site.xml.

```
<property>
 <name>hive.metastore.uris</name>
 <value>thrift://hims1.domain.com,thrift://hims2.domain.com
 </value>
 <description> URI for hive metastore replication
 </description>
</property>
```

By default, hive metastore client treats the first server from the list as the primary host to run metastore service. In case the primary host in unavailable, the client randomly picks up a server to run metastore runs. Note that the metastore relational database should also be enabled for high availability. Similar approach can be followed to setup high availability of WebHcatServer and HiveServer2.

In a security-enabled cluster that requires host authentication, you can allow hive token store by configuring `hive.cluster.delegation.token.store.class` to `org.apache.Hadoop.hive.thrift.DBTokenStore` on all the nodes where metastore service has been running. This setting can be made either through Ambari, if configured, or hive-site.xml.

HiveServer2 and Zookeeper integration

A hiveserver2 instance can be made highly available after integrating with zookeeper. Multiple hiveserver2 instances register themselves with zookeeper. Zookeeper returns a randomly selected instance upon client request. Not just high availability but this also ensures appropriate load balancing.

A hive query getting processed through zookeeper follows below steps.

1. Hive client issues a hive query

2. Hive client connects with zookeeper to receive hiveserver2 details.

   `jdbc:hive2://<zookeeper_hs2_list>;serviceDiscoveryMode=zooKeeper;zooKeeperNamespace=<hs2_namespace>`

3. Zookeeper returns hiveserver2 host and port after random selection

4. Client connects to the host and port

5. Normal query processing steps

Setup HA for Kerberos

Key Distribution Center (KDC) being used can be setup in master-slave mode to server HA requirements of kerberos. Follow the below steps to setup high-availability of Kerberos key distribution center (KDC).

1. Chose a master node as slave KDC and install krb5-server, krb5-libs, krb5-workstation

    ```
    sudo yum install -y krb5-server
    sudo yum install -y krb5-libs
    sudo yum install -y krb5-workstation
    ```

2. Backup krb5.conf on KDC master and create its copy on slave KDC node

3. Backup kdc.conf on slave KDC node

4. Edit kdc.conf on master and slave KDC nodes as below

    ```
    sudo vi /var/kerberos/krb5kdc/kdc.conf
    [kdcdefaults]
    kdc_ports = 88
    kdc_tcp_ports = 88
    [realms]
    DLSEC-SAMPLE.DOMAIN.COM = {
    #master_key_type = aes256-cts
    acl_file = /var/kerberos/krb5kdc/kadm5.acl
    dict_file = /usr/share/dict/words
    admin_keytab = /var/kerberos/krb5kdc/kadm5.keytab
    supported_enctypes = aes256-cts:normal aes128-cts:normal
    des3-hmac-sha1:normal arcfour-hmac:normal des-hmac-
    sha1:normal des-cbc-md5:normal des-cbc-crc:normal
    }

    sudo chmod 655 /var/kerberos/krb5kdc/kdc.conf
    ```

269

5. Update kadm5.acl on master and slave KDC nodes

```
vi /var/kerberos/krb5kdc/kadm5.acl
*/admin@DLSEC-SAMPLE.DOMAIN.COM *
sudo cp /var/kerberos/krb5kdc/kadm5.acl /tmp/
sudo chmod 777 /tmp/kadm5.acl
scp /tmp/kadm5.acl hdp@10.256.39.70:/tmp
scp /tmp/kadm5.acl hdp@10.256.13.2:/tmp
sudo cp /tmp/kadm5.acl /var/kerberos/krb5kdc/kadm5.acl
sudo chmod 655 /var/kerberos/krb5kdc/kadm5.acl
```

6. Update file /var/kerberos/krb5kdc/kpropd.acl on
    Kerberos Master and Slave

```
sudo vi /var/kerberos/krb5kdc/kpropd.acl
host/ip-10-256-38-70.ec2.internal@DLSEC-SAMPLE.DOMAIN.COM
host/ip-10-256-13-2.ec2.internal@DLSEC-SAMPLE.DOMAIN.COM
sudo chmod 655 /var/kerberos/krb5kdc/kpropd.acl
```

7. Install xinetd on Kerberos and initialize kerberos
    internal database on master and slave KDC nodes

```
sudo yum install -y xinetd
sudo kdb5_util create -s
KDC Database MAster Key: xxxxxx
```

8. Create an administrator principal to manage
    Kerberos realm

```
sudo kadmin.local -q "addprinc kdcadmin/admin"
Principal password: xxxxxxx
```

9. Create host keytabs for slave KDC on master KDC

```
kadmin.local
addprinc -randkey host/ip-10-256-38-70.ec2.internal
addprinc -randkey host/ip-10-256-13-2.ec2.internal
```

10. Extract the host key for slave KDC and update the hosts keytab file /etc/krb5.keytab.slave. Copy to slave KDC.

```
-------on Master KDC--------
ktadd -k /etc/krb5.keytab host/ip-10-256-38-70.ec2.
internal
ktadd -k /etc/krb5.keytab host/ip-10-256-13-2.ec2.
internal
sudo chmod 644 /etc/krb5.keytab

-------on Slave KDC--------
scp /etc/krb5.keytab hdp@10.256.13.2:/tmp
sudo cp /tmp/krb5.keytab /etc/krb5.keytab
sudo chmod 644 /etc/krb5.keytab
```

11. Update /etc/services on both KDC hosts

```
sudo vi /etc/services
krb_prop 754/tcp # Kerberos slave propagation
```

12. Configure kpropd on both the KDC nodes in /etc/xinetd.d/krb5_prop

```
sudo vi /etc/xinetd.d/krb5_prop
service krb_prop
{
disable = no
socket_type = stream
protocol = tcp
user = root
wait = no
server = /usr/sbin/kpropd
port = 754
}
```

13. Start KDC and kadmin processes on master KDC

```
sudo systemctl enable krb5kdc
sudo systemctl start krb5kdc
sudo systemctl status krb5kdc
sudo systemctl enable kadmin
sudo systemctl start kadmin
sudo systemctl status kadmin
```

14. Run xinetd as persistent service on both the KDC
hosts

```
sudo systemctl enable xinetd.service
sudo systemctl start xinetd.service
sudo systemctl status xinetd.service
```

15. Replicate KDC database to slave KDC node and start
the slave KDC

```
sudo kdb5_util dump /var/kerberos/krb5kdc/slave_datatrans
sudo kprop -f /var/kerberos/krb5kdc/slave_datatrans ip-
10-256-13-2.ec2.internal
sudo systemctl enable krb5kdc
sudo systemctl start krb5kdc
sudo systemctl status krb5kdc
```

16. Setup a cron to propagate the updates from master
KDC node to slave KDC

NameNode high availability

Until Hadoop 1, the Hadoop cluster used to be controlled through a single
NameNode. if NameNode becomes unavailable due to machine failure,
process corruption, or even planned maintenance, the entire cluster used
to suffer an outage. With Hadoop 2, NameNode can be replicated to a hot

standby NameNode. The standby NameNode remains passive until active NameNode goes down. In case of outages, the architecture supports both manual as well as automatic failover to the standby NameNode.

Architecture

In a Hadoop cluster, the active and standby NameNodes reside on two physically different hosts; one of which actively serves data block requests while other remains in standby mode. Standby NameNode always remains in sync with the state of active NameNode. How this synchronization happens? Let's check out.

Active NameNode logs all system changes (file create/update/delete) that are done to its namespace in editLogs. The edit log contains the incremental system changes after the last purge to fsimage. These edits are synchronously written over to a separate cluster of nodes, known as *journal nodes*. Journal nodes are distributed set of nodes to store the edits. The edits are replicated over the cluster of journal nodes. Standby node scans the new changes from any of the edit replica on journal nodes and applies them to its namespace. This achieves synchronization between active and standby namespaces. In a cluster, minimum of three light weighted nodes can be designated as journal nodes.

In addition to namespaces, another key aspect of high availability architecture is data block coordinates on data nodes. Data nodes are configured for both active and standby nodes. Data nodes communicate heart beat to both the nodes and send block information to both concurrently. This allows standby nodes to be data aware and helps it to assume primary role during fast failover. The architecture diagram shown in Figure 7-2 is for reference.

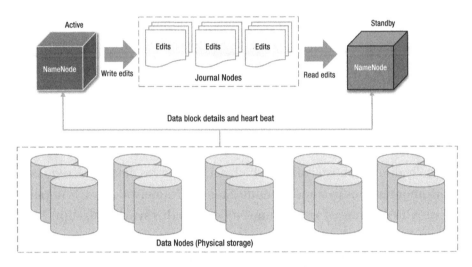

Figure 7-2. *NameNode high availability architecture*

Keep in mind that active-active configuration of name nodes will not be possible as it may lead to split-brain situations. Journal nodes allow only one writer process at a time. If the NameNodes and journal are not in sync, it may cause a huge risk to data precision and availability.

Below is the sample configuration in hdfs-site.xml for high availability of NameNodes.

```
---------Logical name for new Nameservice---------
<property>
  <name>dfs.nameservices</name>
  <value>dl_clustr</value>
</property>

---------Configure list of NameNode identifiers in each
nameservice---------
<property>
  <name>dfs.ha.namenodes.dl_clustr</name>
  <value>nn_active,nn_stby</value>
</property>
```

```
---------RPC address for each NameNode---------
<property>
  <name>dfs.namenode.rpc-address.dl_clustr.nn_active</name>
  <value>dlmc1.machine.com:8020</value>
</property>
<property>
  <name>dfs.namenode.rpc-address.dl_clustr.nn_stby</name>
  <value>dlmc2.machine.com:8020</value>
</property>

---------HTTP address for each NameNode---------
<property>
  <name>dfs.namenode.http-address.dl_clustr.nn_active</name>
  <value>dlmc1.machine.com:50070</value>
</property>
<property>
  <name>dfs.namenode.http-address.dl_clustr.nn_stby</name>
  <value>dlmc2.machine.com:50070</value>
</property>

---------JN URI where the edits would be written and read by
NameNodes-----
<property>
  <name>dfs.namenode.shared.edits.dir</name>
  <value>qjournal://jnode1.machine.com:8485;jnode2.machine.
com:8485;
jnode3.machine.com:8485/dl_clustr</value>
</property>

---------JN local directory where edits could be
persisted---------
<property>
  <name>dfs.journalnode.edits.dir</name>
```

```
  <value>/dfs/journal/localdata</value>
</property>

---------Java class to ping Active NameNode---------
<property>
  <name>dfs.client.failover.proxy.provider.dl_clustr</name>
  <value>org.apache.Hadoop.hdfs.server.namenode.ha.ConfiguredFa
iloverProxyProvider
</value>
</property>

---------HA fencing configuration using shell method---------
<property>
  <name>dfs.ha.fencing.methods</name>
  <value>shell(/path/to/my/script.sh arg1 arg2 ...)</value>
</property>
```

Design considerations

Below are the factors that play a vital role while panning high availability of Hadoop NameNodes.

1. Generally, it is a good practice to maintain odd number of journal nodes to survive maximum failures.

2. Currently, only two nodes, can be provisioned to achieve high availability. Out of the two configured NameNodes, whichever starts first is considered active.

3. Guidelines for HA configuration

    a. All nameservices should be added to dfs. nameservices. This list should include all the nameservices which are used for NameNode federation.

b. Maximum of two namenodes associated with a nameservice should be added for each nameservice in `dfs.ha.namenodes.[nameservice ID]`

c. Add group of journal nodes in `dfs.namenode.shared.edits.dir` for storing shared edits.

d. If you are not using custom class to determine active NameNode, set `dfs.client.failover.proxy.provider.[nameservice ID]` to `org.apache.Hadoop.hdfs.server.namenode.ha.ConfiguredFailoverProxyProvider`.

e. Quorum Journal Manger prevents multiple NameNodes to write edits on to journal nodes.

4. Fencing active NameNode inhibits split brain situation by restraining two NameNodes from writing edits on journal nodes at the same time. It is always a good practice to fence NameNode even when using quorum journal manager (QJM). During failover, fencing ensures that the dying NameNode doesn't serve any few read requests before shutting down completely. Hadoop provides two configurable fencing methods namely, *shell* and *sshfence*.

a. Custom fencing logic can be specified in `org.apache.Hadoop.ha.NodeFencer`

b. *shell* enables users to run a shell command in lieu of what a usual NameNode operation could be. The shell command may not be business relevant but will proxy the data node request for a NameNode.

   i. Connection timeout can be configured using `ssh.connect-timeout`. Connection timeout indicates failed fencing.

    c.   *sshfence* allows ssh to a target node and kills the NameNode process using fuser. Passphrase key to the target node must be owned by hdfs user and should be available in `dfs.ha.fencing.ssh.private-key-files`.

    d.   The result of fencing operation must be a success; else failover to standby node might not happen

5.   Manual failover – on the candidate (target) NameNode, execute

```
hdfs haadmin /
-failover /
--forcefence /
--forceactive <serviceId> <namenodeId>
```

6.   Automatic failover configuration – Done through zookeeper quorum and ZKFailoverController (ZKFC) process.

    a.   Apache zookeeper is a high availability service in Hadoop cluster that monitors the cluster component for failures. Automatic failover mechanism requires detection of the event when active NameNode fails and election of next active node. The zookeeper service stays in a live session with all NameNodes. As soon as the session expires when active NameNode undergoes failure, zookeeper triggers a failover notification. Simultaneously, the successor target node acquires an exclusive lock on zookeeper to indicate it as the next primary NameNode.

   b. ZKFC is a zookeeper client which is responsible for health monitoring of NameNode by pinging and managing the session with the active NameNode. During active namenode election, ZKFC helps healthy NameNode in acquiring lock.

```
---------Automatic failover configuration for a
nameservice-------
<property>
  <name>dfs.ha.automatic-failover.enabled.[NameService_
  ID]</name>
  <value>true</value>
</property>

---------List of hosts running zookeeper
service---------
<property>
  <name>ha.zookeeper.quorum</name>
  <value>zookee1.machine.com:2181,zookee2.machine.
  com:2181</value>
</property>
```

   c. Zookeeper security – setup zookeeper authentication and ACL for zookeeper access in core-site.xml

   d. Use the following command can also be used to query the HA state of a NameNode

```
hdfs haadmin -getServiceState
```

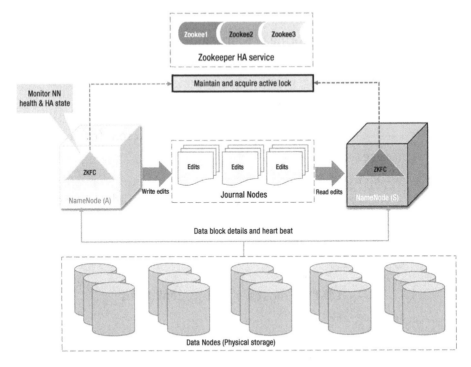

Figure 7-3. *Integrate NameNode HA services with zookeeper for continuous monitoring and proactive alerting*

Data Center disaster recovery strategies

Until now, what we focused on was the high availability of Hadoop components. Practically, in an enterprise big data lake, what we need to prevent from losing is data and analytical models. Data lake may lose data in the events of logical error or hardware failure. While logical errors lie in the purview of developers and analysts, hardware failure can be further classified as a function of risk and cost. With cost playing an adjudicator, incidents like hardware failure due to system crashes, media failure, or node failure have high risk but low-cost impact. Therefore, they can be prevented by handling failure situations are the platform level. However, there could be occasions where entire data lake becomes inaccessible due to hardware failures, mass power outage or network failure or site shuts down under civilian circumstances. Disaster recovery targets those instances where a data center can take over a primary data center within permissible service outage thresholds.

For data management professionals, disaster prevention and recovery is not a new term. While the objective remains same, the disaster prevention strategy does changes based on service level agreements, and platform architectures. In this section, we are going to discuss disaster prevention and recovery strategies pertaining to Hadoop data lake.

Why to replicate a data lake? Setting up a data lake demands huge efforts in cost, planning, defining technical design and architecture, and streamlining daily operations. Standing up another infrastructure will shoot up the capital investment as well as operational overhead. Therefore, before moving on to planning phase, one must have a strong use-case and clear objective behind setting up a replica (s). There are two parameters that justify a replication exercise: *availability* and *recoverability*. You tend to achieve availability by having redundant or additional supply of resources for tolerating a fault or an outage without (or minimally accepted) incurring any loss to the business. Recoverability can be achieved through an alternate standby site that holds as-of-outage state of data and can be quickly restored. High availability can be achieved by planning high availability of member components of a site. Recoverability addresses bigger concern when entire site has to be failed over to its standby. Therefore, availability happens to be the subset of recoverability. An active-passive site could be an optimal approach that achieves recoverability and availability. An active-active setup attains the state of *nirvana* by enabling active replicas to the business users, while both treat each other as standby.

With cloud service models decently prevalent into IT these days, most of the cloud vendors promise high availability (as high as ~99.999%) for cloud hosted applications. Recoverability may vary by nature and service level agreements of applications. For example, a "customer feedback" application can compromise an outage of couple of hours, but a "sales" application cannot. A business-critical application needs immediate restore to its standby site in order to prevent business disruption.

Although cost of investment becomes a driving consideration, but the organizations must determine a calculated measure of both the factors, while justifying the proposal for a standby site. Companies with global footprints and reginal governance laws, are forced to have an active replica for uninterrupted analytics. In a similar scenario, a data center located in a place which is frequent hit by natural hazards, might be looking for a passive replica for recoverability purposes.

Disaster recovery factors

At a high level, disaster recovery strategy involves a backup site and switchover strategy. The nature of backup for disaster recovery is slightly different as the expectation from disaster recovery is to cope up from critical incidents. Let us list down the factors that play their part in formulating an efficient disaster recovery strategy.

1. Understand data sources and data awareness – While setting up a disaster recovery site, it is always a better idea to understand ingredients of data lake. How critical are the system of records? Where do the source system exist? What is the impact if a data mirror layer is lost?

2. Copying versus mirroring – Backup mode is an essential parameter of restoration exercise from disaster recovery site. Mirror images restore faster than backup copies.

3. Backup frequency – The data change factor and service level agreements determine the frequency at which data flows into the disaster recovery site.

Disaster recovery approaches

Let us start looking at the building blocks of disaster recovery strategies. Keeping the above considerations in mind, there could be two possible approaches to start with.

1. Dual path ingestion or *Teeing* – Under this method, all distinct source systems follow a T-like two-way ingestion pipeline and push data in production as well as standby (or replica) data lake. Ingestion pipeline may or may not be the same as between data source and primary data lake. Though it can be reused to throw data into standby site, but parameters like scalability, cost, and performance need to be factored in before channeling it for standby purposes. Nevertheless, modern commercial tools give flexibility to enable two-way replication at different frequencies.

   The model shown in Figure 7-4 invites more arguments within the architect community than benefits. However, in a typical "data as an asset" world, it makes sense to ingest just the mirror layer. Consumption layer can either be built in parallel or whenever required by running business models to consume mirror layer data and produces analytical insights.

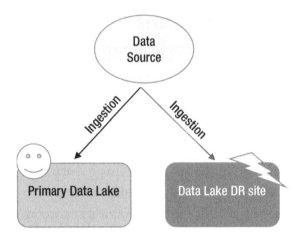

Figure 7-4. *T-ingestion approach to ingest data in production as well as replica cluster*

There are quite a few challenges with the parallel data
ingest approach.

a. Parallel data ingestion pipeline will shoot up the
resource consumption. Ingestion framework
configuration like bandwidth, network resources
needs to be re-evaluated and most possibly,
stretched out. A load balancer would be desirable
in the implementation model.

b. In an active-passive setup, active site is presumed
to be critical. If parallel writes are run in
asynchronous mode, maintaining standby site
as actual as primary becomes an operational
overhead. The chances of standby site getting
diverged from primary are high.

c. Only source data can be ingested in parallel to
both sites and not the processed data (stage or
consumption layer). Data processing models may

run on standby sites; however, data consistency cannot be guaranteed. In case, business users agree on using read-only standby site for ad-hoc exercises, data operations must employ regular checks and balances in place to safeguard the sanity levels.

2. Data Center replication or *Copying* – This is a most prominent approach which is practiced quite often while planning disaster prevention measures of an enterprise data lake. With this technique, data from source systems gets ingested into primary data lake only. From the active data lake, data moves to its DR site as shown in Figure 7-5.

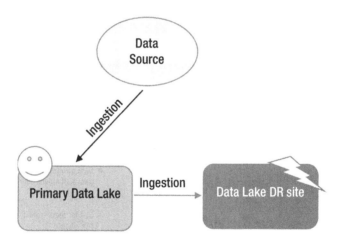

Figure 7-5. *Copy approach to replicate data from production to replica*

This approach provides the flexibility of replicating mirror as well as consumption layer to the standby site. In case of disaster, standby can quickly resume the role of active data lake without prolonged outage cycles. In addition, since no processing is required to run on replica cluster, it can be used for ad-hoc analytics and visualization.

Contrary to the previous approach, it requires less resources and thus, source-to-mirror ingestion framework remains untouched. At the same time, it puts the pressure on active data lake to replicate data to the standby site.

While planning for large data from the production data lake to its replica site, consider the below points.

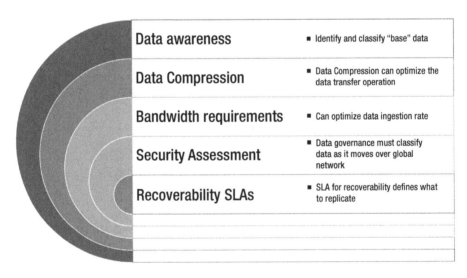

***Figure 7-6.** Key considerations when replicate large volume data from production to replica clusters*

Data replication strategies

The dual path approach for high availability is a straightforward one. We will stick our focus on the second approach that requires development of new ingestion framework between active data lake and its standby site. Factors that impact the ingestion pipeline are nearly same – data volume, frequency, impact on data lake, resource management, and batched ingestion versus change data capture. In this section, we will walkthrough data replication strategies that can be practiced on a production data lake site. These techniques form the base of many commercial and open source replication tools as well.

1. DistCp – DistCp or Distribution Copy is one of the most common copy solutions for Hadoop file systems within the same data center or remote data centers. Under the hoods, it uses mapreduce for data distribution and recovery. It translates list of directories and files under a namespace into map tasks and taskTrackers copy them over to target namespace.

   Although the utility usage is pretty simple, but the approach has some obvious limitations. The utility being a mapreduce operation, may consume few map slots that may impact the business operations in data lake. In addition, since each datanode on the source site should have write access on target sites, the communication pattern between the two clusters is SN*TN [SN is the count of source data nodes, TN is the count of target data nodes]. In case the communication channel is not setup one-on-one between data nodes, data replication from source to target may get impacted.

Another key consideration of distcp usage is Hadoop version on source and target. With hdfs:// connection protocol, the source and target versions must be same. To switch on version independent transfer between source and target sites, enable data transfer over HTTP by using wedhdfs://. Another method of enabling HTTP-based transfer is using httpfs:// protocol, which uses HTTPfs proxy daemon for cluster communication. However, keep in mind that both webhdfs:// and httpfs:// need to be configured manually and are relatively slower than native hdfs:// connection.

2. HDFS Snapshots – snapshots represent state of data lake at a point in time. HDFS snapshots can be build an as-is image of data lake. In addition, they can be used to stitch data during accidental losses.

3. Hive metastore replication – Hive supports metastore replication to other clusters with simple configuration in hdfs-site.xml file. Although custom replication frameworks are possible, but by default, system uses `org.apache.hive.hcatalog. api.repl.exim.EximReplicationTaskFactory` implementation for data capture, movement, and ingestion commands.

4. Kafka mirror maker – Apache Kafka service that acts as a consumer in active Kafka cluster and producer to standby Kafka cluster.

With global footprints becoming more routine, companies strive for data availability for globally situated teams and for this, they require a robust data replication solution that can encompass geographically located sites and carries the ability to handle voluminous data sets in near real-time (real-time would be incredibly welcomed though!).

With hive as a data warehousing realtor in the data lake world, it makes a lot of sense to setup change integration between two regional data lakes. The change integration layer accomplishes two tasks. First, it captures and emits the changes to the standby site. Second, it instructs or mimics the in-built writer process to ensure timely merging of changes. With a stable integration layer, we can enable multi-directional replication and setup a multi-site data lake. The implementation of integration layer demarcates availability versus recoverability parameters of data lake. Let us go through design considerations of active-passive and active-active models of setting up high availability sites.

Active-passive data center replication

In an active-passive setup, the standby data lake site remains in passive state until the event of disaster on primary site. During disaster, the primary site suffers outage and standby presumes the role of the active site. Once the originally active site comes up, it operates in standby mode.

Since the changes must flow unidirectionally, replication can be achieved via periodic synchronization through DISTCP. A custom integration layer can also help in batching the changes and pushing them over to the secondary site.

Active-passive approach appears more like backing up the site for future recoverability, rather than available. It is read-only replica of primary which, due to periodic synchronization, may become inconsistent. As a result, it becomes an operational overhead to catch up the data lags and bring in pace with the primary.

Active-active data center replication

Not so long ago, multi-site live data lakes were a concept unexplored, but the drastic rise in expectations, audience, and agreements has kept things moving since last couple of years. With data democratization becoming a thing and evolution of "citizen" data scientists out of blogs and books, multi-site "live" data lakes have started hitting the practice.

Active-active replication allows you to ingest data from any regionally located site. Fresh data will be replicated to all other sites in the network. Although it challenges data governance to its limits, but you can ingest anywhere, and analyze anywhere. It enables maximum resource utilization within a site, brings data consistency, offers disaster coverage, shares workload, and fences regionally located users on a democratized platform.

High level architecture diagram of active-active replication between regional sites is shown in Figure 7-7. Note the "Change Coordination Engine" component. It is a distributed coordination engine which is responsible for emitting changes across all data lake sites. Changes are nothing but any *write* request from the client. The change coordination engine serves two purposes:

1. Synchronously replicates metadata across data centers

2. Maintains order of transactions and replicates data asynchronously

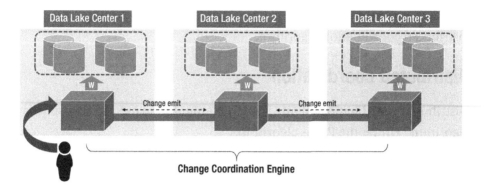

Figure 7-7. *Active-active replication across data lake sites using distributed coordination engine*

The idealistic architecture depicted in Figure 7-7 is implemented by WANDisco Fusion. In the next couple of sections, we are going to highlight the active-active replication capabilities of the product.

WANDisco Fusion

WANDisco Fusion provides an active-active replication technology across data centers. It maintains an illusionary server system, cutting across the data lake sites and can perform at LAN speed over widely distributed environment. Its proxy server architecture replicates every change for selected HDFS folders.

WANDisco replicates data in block and sub-block increments and does not require a file to be fully written and closed before replication. Under the layers, it uses a highly efficient and fault tolerant coordination engine that achieves ordered transaction management in a distributed environment. It deploys a quorum-based configurable approach to freeze the order of transactions. A quorum is a subset group of participating nodes of the coordination cluster.

Distributed coordination is decentralized engine that manages the transactions processing across all sites. A dedicated central coordinator can also fit the bill but exposes the risk of single point of failure and becomes scalability bottleneck at times. Let us briefly understand the distributed coordination algorithm.

Distributed coordination

Distributed coordination engine is an implementation of Paxos consensus algorithm.[1] Under the Paxos algorithm,[2]

> *A replicated state machine is installed with each node in a distributed system. The replicated state machines then function as peers to deliver a cooperative approach to transaction management that ensures the same transaction order at every node.*

Distributed coordination engine has an agent installed on nodes of a distributed cluster, which forms a virtual namespace. While virtual namespace of fusion nodes is consistent with respect to events, the agents or nodes can attain the role of either *proposer*, or *learner*, or an *acceptor*. You can understand these roles as the phases of the process for reaching consensus on an active transaction. *Proposer* phase marks the election of a node from the virtual namespace. *Broadcast* phase submits the "transaction" proposal to other fusion nodes in the cluster for consensus. Under *accept* phase, quorum of nodes accepts the proposal emitted by the proposer; thereby reaching the consensus and establishing the order of global sequence of events. Once consensus is reached, the proposer broadcasts the commit messages to all fusion nodes to indicate "go-ahead" with the transaction. Keep in mind that only cluster writes are coordinated by the engine and not he reads. The flow and function of fusion node roles are described in Figure 7-8.

---

[1]The Paxos algorithm was designed by Leslie Lamport to provide a fault tolerant and decentralized framework for enabling active-active replication.

[2]Refer - https://www.wandisco.com/assets/blt1d792cb4d9252692/WANdisco_DConE_White_Paper.pdf

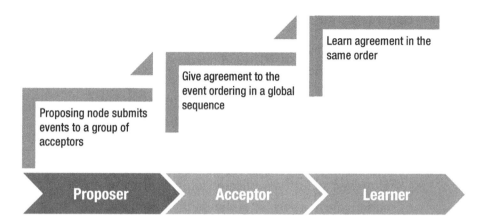

Figure 7-8. *Roles and phases of a distributed coordination engine*

Design considerations

Below are the key characteristics and factors that can help in understanding the replication mechanism better.

1. No changes in the working model of existing Hadoop components. NameNode, DataNode, and MapReduce continue to operate as usual.

2. Fusion employs Inter Hadoop Communication Service with the storage systems like Isilon, Hadoop, MapR, and Amazon S3. It also supports heterogeneous storage zones that could possibly execute storage tiering policy for data lifecycle management.

3. Performance and fault tolerance is achieved through the concept of quorum-based agreement.

4. Consistency model

    a. Fusion nodes coordinate to preserve the order of deterministic updates

    b. Proposal lifecycle is exercised on global sequence of agreements. State of the folders that are set for replication may have different states due to the difference in agreement consumption rate. However, all fusion replicated folders carry the same state at a specific global sequence number.

    c. Election of a proposer – A node issues a fresh proposal with its sequence number higher than the last that it was aware of. Upon proposal broadcasting, if quorum of nodes reply affirming the high value of proposal sequence number, the issuer node is elected as the leader. Once the leader of a proposal is elected, the contending and pretending coordinators cannot proceed until its consensus settlement.

    d. If multiple nodes pretend to be coordinators for the same proposal, the algorithm restricts their choice of value selection through ordering.

5. WANDisco Fusion supports replication of selective data across data lake sites – Complex replication use cases like regional data governance laws, data restrictions, can be configured.

Figure 7-9 outlines the flow of a transaction proposal from Distribution Coordination Engine's local instance to consensus state.

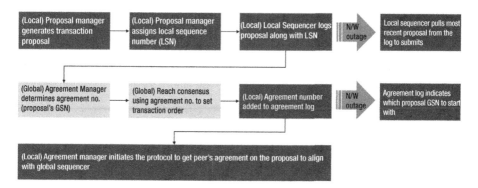

Figure 7-9. *Achieving consensus through quorum-based approach with Coordination engine*

Conclusion

Enterprise data lake is no less than a sea of information. With asset comes the responsibility and challenges. Availability and scalability are two of the top mission level considerations of data lake strategy. In this chapter, we discussed how Hadoop natively handles scalability through NameNode federation and shares the load with other ones. From the high availability standpoint, Hadoop has no defined strategy but as a user, guidelines can be laid down for business continuity and disaster prevention.

The next chapter is going to talk about how to achieve operational success in a data lake. A full-blown data lake demands a body who not just monitors the health of data lake and send out notifications, but also acts as the custodian to platform upgrades, data lake scalability, and documents and deployments.

CHAPTER 8

Managing Data Lake Operations

"Without big data analytics, companies are blind and deaf, wandering out onto the web like deer on a freeway."

— Geoffrey Moore, Author, "Crossing The Chasm"

By now, the readers would have got a fair understanding of data analytics in real world and how data lake caters to the needs of data analytics. All organizational data assets converge under one hood and conceptualize complex data sets into a full-blown data lake. It is essential to understand how to strive for a healthy, stable, and secure data lake. How an organization tackles security, stability, and availability challenges to ensure data lake remains live and adheres to compliance guidelines?

As they say, it is easy to create, but difficult to sustain. Managing a production cluster can get as complex as it can be. Lack of platform understanding, application SLAs and an efficient monitoring framework can bring opacity in data lake operations. As soon as data lake ecosystem stabilizes and becomes operational, it becomes critical to attend its pressing requirements as listed below. Keep in mind that these requirements, eventually formulate into key processes of operations desk.

© Saurabh Gupta, Venkata Giri 2018
S. Gupta and V. Giri, *Practical Enterprise Data Lake Insights*,
https://doi.org/10.1007/978-1-4842-3522-5_8

The list compiles the standard operations checklist, while there could be more granular and precise tasks as agreed during regular handshakes between data management and operations.

1. Gatekeeper of data lake platform – Operations remain the owner of production environments. Liaison with internal IT teams, data governance council and data lake development to be aware of data lake objectives and its primary stakeholders.

2. Support data lake availability in line with SLAs defined by downstream consumers – Introduction of a layer for proactive monitoring and alerting keeps a constant check on platform availability.

3. Provide operational intelligence and publish metrics to highlight areas of risk – Setup regular rhythm to perform incident analysis and health checkup of the platform and business application. Publish key metrics that highlight availability numbers, issue trends, and application readiness scorecard.

4. Integrated support through regular communication – Communication holds the key to quick turnaround on issue resolutions. Over the time, it polishes the ability to sense risk swiftly and display smart acts during remedy actions. Bridge development and stakeholders to get their feedback and issues, if any.

5. Be agile and nurture the culture of "continuous integration" and streamline deployment process – Encourage continuous integration, continuous delivery and continuous deployment principles to smoothen delivery pipeline.

This chapter will primarily focus on the principles of monitoring architecture that will allow administrators to incubate operational excellence with data lake ecosystem. It will give them an insight of how compelling functions of data lake can be optimized and made robust to enhance transparency and accountability. Within the scope of the chapter, it will not deep dive into Hadoop operational structure.

Monitoring Architecture

Data lake operations teams are often confronted with questions from application users like ones shown in Figure 8-1.

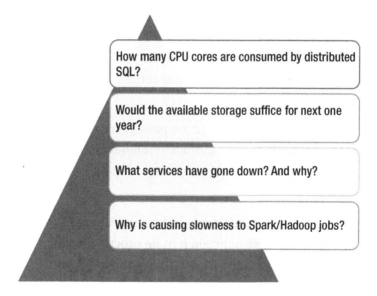

Figure 8-1. *Questions that data lake operations often encounter*

A good monitoring architecture not only empowers administrators with cluster monitoring capabilities, but also with performance measurement of applications. Basic cluster monitoring only tells you what has failed but does not answer why it has failed.

Hadoop metrics architecture

For any system, trace logs, audit logs, or any sort of statistic information is important to understand system's health and behavior. This informative statistical data which reveals system's state is known as *metrics*. Metric is generated by daemons that run on components of a system and can be used to monitor component's health, tuning, and troubleshooting. In Hadoop, there are plenty of metrics that are generated by default. However, for additional metrics, separate agents can be installed on specific components.

In a Hadoop based data lake ecosystem, metrics are grouped under focused contexts. Metric is a line level information in a context. A context can be configured with a plug-in that operates on a particular component of Hadoop, thereby generating all metrics grouped under the context. Each metric is tagged with the hostname so as to differentiate from rest of the metrics. Let us see what are the critical contexts available in Hadoop metric architecture.[1]

1. *jvm* – It groups metrics that are published by java processes and reveal information related to memory used and committed, thread details, and process trace logs.

2. *rpc* – The metrics under *rpc* context are generated during handshake between remote procedure calls and the host. It could be related to authentication, data volume transfers, RetryCache, and open connections. Another related context *rpcdetailed* generates additional metric for rpc methods. It is not included in the *rpc* metric record.

---

[1]https://Hadoop.apache.org/docs/r2.7.2/Hadoop-project-dist/Hadoop-common/Metrics.html

3. *dfs* – This context contains metric for NameNode, FSNamesystem, JournalNode, and DataNode. One of the most important context that collects metric for namenode operations, capacity, journal sync, and data node operations.

4. *yarn* – Yarn context captures metric for NodeManager, cluster, and queue manager.

5. *mapred* – This context generates metrics through MapReduce daemons. It contains metrics related to jobtrackers and tasktrackers.

In addition to the above contexts, there could be contexts to give additional metrics like ugi, *metricsystem,* and *default* contexts. The ugi context groups metric related to users and groups like successful and failed kerberos logins, and group resolutions. The metricsystem context reveals metrics about the metric sources.

Identification of source components

This is the first step towards developing a good monitoring framework. An organization's data-lake could be built using any technology stack (Hadoop, Cassandra etc), Hadoop being the most notable. All necessary services must be identified that are running on technology stack platform. For instance, some of the essential Hadoop components that should be monitored and whose metrics are to be collected are listed as below.

YARN metrics

Active nodes or Lost nodes list - This metric should give us the count of nodes operating without any problem. Nodes can lose contact with a resource manager for variety of reasons ranging from network issues to lack of hardware resources (cpu, memory, etc.).

If a NodeManager (NM) is unable to reach a ResourceManager (RM) for a given timeout threshold, that NM will be marked as lost and its resources unavailable for the cluster. We should take action once the node is marked "lost".

Total amount of memory allocated - This metric gives a high-level overview of cluster memory usage. If we are frequently hitting close to the cluster capacity, it is good time to investigate which jobs could be consuming and enough attention to be paid to tune such jobs. The other option is to add NodeManager and/or increase the amount of memory reserved for YARN.

MapReduce metrics

To optimize or find bottlenecks in their application, developers should keep track of

1. Number of failed maps

2. Number of failed reducers

3. Data-locality counters – Hadoop also exposes a set of metrics that tells how closely does a job create a map to its data. If many map tasks are created on nodes where the data is not available locally, it gives a good indication of degraded performance.

HDFS

There are multiple parameters that give a good indication of health of HDFS such as:

1. Total count of files

2. underReplicatedBlocks

3. StaleDataNodes

4.   CapacityRemaining

5.   BlocksTotal: Current number of allocated blocks in
     the system

For an exhaustive list of metrics exposed by each service in HDFS
(NameNode and DataNode), please refer to apache documentation:
`https://Hadoop.apache.org/docs/r2.6.0/Hadoop-project-dist/`
`Hadoop-common/Metrics.html`

The data lake in your organization could end up with multiple
ecosystem components like HBase, Hive, Spark, and other SQL-on-
Hadoop technologies like Apache Drill, Presto etc. For each such
component, you must include certain necessary metrics that will help
monitor them in the next stage we will be seeing soon.

The key point to note here is that almost every Hadoop ecosystem
component exposes its metrics via a JMX port that we can connect and
track those metrics for monitoring and performance measurement.
Apart from metrics which will help us in monitoring and measurement of
performance, it is important to collect the logs generated by every Hadoop
component.

Metric collection tools

One of the important tools that does all the hard work of collecting the
metrics is CollectD. A daemon collects system performance metrics
periodically and provides mechanisms to store the values in a variety of
ways or to send to the next stage where data is aggregated by other tools
and eventually ready for consumption by a visualization tool. It runs on
every node in the cluster.

Note – Enable the following default plugins which help collect the relevant metrics:

```
[root@vm75-132 ~]# cat /etc/collectd.conf | grep '^Load'
LoadPlugin syslog

LoadPlugin cpu

LoadPlugin df

LoadPlugin disk

LoadPlugin interface

LoadPlugin load

LoadPlugin memory

LoadPlugin network

LoadPlugin swap
```

If you come from a traditional database background these are very similar to the agent processes, ex: Oracle agent in Oracle world or Nagios agents that runs on every node send the metrics to a Manager process. In fact, they are known as "collection agents".

The other well-known enterprise-ready component is Ganglia. It has been in use at some of the biggest cloud infrastructures and is supported by an active community.

Similar to metrics collection agents, other Log collection agents are FluentD and Logstash.

FluentD is a daemon process that runs on each node to collect and parse the logs. These logs can then be sent to data stores like OpenTSDB or backend stores like ElasticSearch. FluentD sees through wide adoption these days as it is known for its built-in reliability and less memory usage.

Logstash is mostly known as being part of ELK stack (ElasticSearch-Logstash-Kibana). Users should make a good comparison of how each tool works and make an informed decision to go for a particular tool in favor the other.

Metrics and log storage

OpenTSDB is a distributed time-series database running on top of HBase. It can be used to store and aggregate the logs and metrics received by agents like Fluentd and collectd. The schema of openTSDB is highly optimized for fast aggregations of a time-series data. Apart from being distributed in nature, it can store huge amounts of data in fine-grained nature.

But the implementation of openTSDB is a bit more complex than installing and configuring systems like Graphite. Nevertheless, if you plan to deploy HBase as one of the ecosystem components in your data lake, you should seriously consider it as a datastore for your logs and metrics.

ElasticSearch(ES) is an open-source distributed full text search and analytics engine that can be one of your Hadoop components in a data lake. Complex search facilities and analyses of logs are one of the best use-cases of ElasticSearch. It is very useful if you want to analyze and mine the data to look for trends, statistics, summarizations, or anomalies. You can also use Logstash, part of the ELK stack to collect, aggregate, and parse your data, and then have Logstash feed this data into Elasticsearch.

It indexes the logs received via collection agents so that they can be easily accessed and searchable. For instance, FluentD uses round robin method when writing the logs to ES. If for some reason, one of the nodes is not available, fluentd can failover to one of the surviving ES nodes. You don't need additional configuration for setting failover for fluentd to work with ES. But it is recommended to have a minimum of 3 nodes for running ElasticSearch. You must size the cluster with one of more nodes as per your requirements.

By default, it indexes the data (received logs and metrics) for 2 days. But can be easily configured for higher retention period. ES is also known to be memory intensive and uses a minimum of 2Gb RAM heap size. This can again be configurable as per your needs. With all the aggregated data placed in Elasticsearch, you can search for any combination of nodes, services, or message severity levels that you want to monitor and further develop alerting and visual analytics on top of this.

The repo location for ElasticSearch may not be available for CentOS/ RHEL systems by default. Therefore, import the Public GPG key of ElasticSearch into rpm and then install it as per the below steps.

```
//Import GPG Key//

[root@vm75-132 ~]# rpm --import http://packages.elastic.co/
GPG-KEY-elasticsearch
[root@vm75-132 ~]#

//Add a repo location//

[root@vm75-132 ~]# vi /etc/yum.repos.d/elasticsearch.repo
[root@vm75-132 ~]# cat /etc/yum.repos.d/elasticsearch.repo
[elasticsearch-2.x]

name=Elasticsearch repository for 2.x packages
baseurl=http://packages.elastic.co/elasticsearch/2.x/centos
gpgcheck=1
gpgkey=http://packages.elastic.co/GPG-KEY-elasticsearch
enabled=1

//Install ElasticSearch and start the service//

[root@vm75-132 ~]# yum -y install elasticsearch
[root@vm75-132 ~]# systemctl start elasticsearch
```

There is no explicit configuration required for ElasticSearch here. The only items to be edited are NodeName and cluster name. Modify them as per your cluster/node names.

Logs and Metrics visualization

This layer has seen some rapid advancements recently. There are some awesome front-end tools and libraries with rich feature sets supported by both OpenTSDB and Elasticsearch.

Graphite has been very popular in this category for some time. But like all things with Hadoop, there are other tools like Grafana and Kibana that are very active these days with an active user base.

For the scope of this article, we will only provide an overview of capabilities of Grafana and Kibana. Grafana uses REST API to access metrics data from OpenTSDB. Using a single instance of Grafana, users can build custom dashboards or use sample dashboards to visualize the metric. It also supports many different datastores like Elasticsearch and Graphite.

Grafana has the ability to combine data from multiple data sources and display them in a single dashboard. Each data source is closely tied to a single pane/frame in the dashboard. Since there are a variety of backends, the query language to be used is different as well.

Figure 8-2 shows a Grafana console showing Node level CPU, Memory, Network, and Swap usage which are the some of the indicators of load on the system.

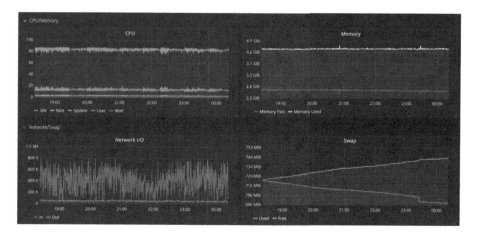

Figure 8-2. *Grafana console showing key system metrics*

You can also tweak the console context to show running metrics (Figure 8-3).

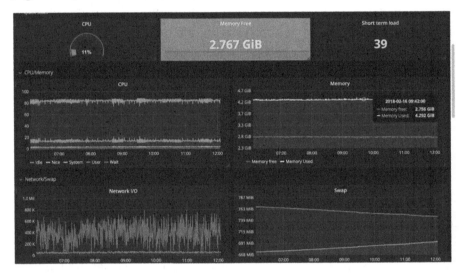

Figure 8-3. *Grafana console context can be modified to show current metrics*

Kibana

Kibana is another visualization platform that runs on top of Elasticsearch. It uses REST API to access and search the logs available in Elasticsearch. Using a single instance of Kibana, users can create visualizations and dashboards to analyze their logs.

Kibana connects to a single Elasticsearch node to read logs. In the event that Kibana is unable to read logs due to the failure of an Elasticsearch node, we have to configure Kibana to connect to an available Elasticsearch node. In case, you want to run Kibana when the configured ES node is down, you can do this by running a Coordinating only ES node on the same node where Kibana is running on the cluster. These coordinators act as load balancers and distribute the incoming connection requests from Kibana to other ES nodes, gather the results and return them back to Kibana for visual representation.

Apache Ambari

Apache Ambari is an open source platform to manage and monitor a Hadoop cluster. The capabilities are not only restricted to operations, but it also enables provisioning and controls security framework of the cluster. Architecturally, it is no different from other contemporary management applications. The server works with agents per component deployed on it to receive back their state as metrics and logs.

Ambari offers the below features for platform monitoring and operation management:

1. Ambari alerts – Apache Ambari raises pre-defined and centrally-managed alerts. Alerts can be modified to control threshold, recipients, frequency, and notification. Alerts offer complete view of cluster health. Ambari offers variety of configurations to customize alerting mechanism.

2. Ambari metrics – Consists of metrics collector, metric monitor, and grafana. Grafana includes multiple pre-build dashboards for visualizing key metrics. Below is the flow of metric flow from its generation to visualization.

   a. Metric monitor publishes system metrics to the collector

   b. Sink pushes Hadoop metrics to collector

   c. Metric collector aggregates the metrics

   d. Metrics displayed over Ambari UI through REST API

   e. With Ambari 2.2, grafana serves as native interface for metrics

      i. Dashboards for HDFS home, namenode, data node, YARN, applications, job history, etc

3. Kerberos – Ambari enables wizard-driven Kerberos administration from the interface. One can create kerberos principals and keytabs, distribute keytabs, and undertake cluster configuration tasks.

4. Role-based access control – access to the cluster can be controlled through roles and permissions. For example, a role *admin* may perform the role of a service as well as cluster administrator.

5. Log Search – Apache Solr enables the component logs to be rapidly searched without any hassle from within Ambari. Search criteria can consist of keywords, time range or logging level.

6. Extensibility – While Ambari stays agile, it can be extended to add or modify a service for custom environments. Ambari interface views can be extended to modify web components.

7. SmartSense – It is an auto diagnostic tool that collects incident information, creates a "bundle," and uploads it to the Hortonworks support. This expedites the incident resolution with reduced turn-around time. Furthermore, it analyzes the bundle and produces recommendations for each cluster. Recommendations aim at reducing operational issues and better cluster performance.

Figure 8-4 branches out the capabilities of Apache Ambari.

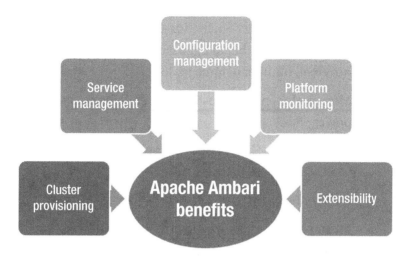

Figure 8-4. *Apache Ambari benefits*

Data lake operationalization

Integrally, the journey of a data lake from ideation to planning, provisioning, and operationalization requires subtle business acumen and organizational vision. Keeping aside the efforts invested in ground work, leadership consensus, and alignment to organizational strategy, some of the critical stages of a data lake are capacity planning, provisioning, monitoring, and security.

The below list discusses design considerations that could be able to run a data lake. More than the considerations, the list can be treated as best practices to ensure platform stability of ecosystem.

1. *Cluster planning* – A Hadoop-based data lake can be provisioned either on an on-premise site or in the cloud. Both on-premise and cloud have gone through several debates of pros and cons. While cloud is meant to provide stability, availability, and better return on investment (ROI), on-premise gives the flexibility to be ductile as per requirement and use case.

2. *Cluster design*

    a.   Chose more number of light-weight nodes and not small number of large nodes.

    b.   More nodes enhance resilience, parallelism, and power

    c.   Less number of large nodes expose several issues like longer recovery time

3. *Component layout*

    a.   Master components should be distributed across the rack to mitigate the risk

    b.   Worker components should be identical across worker nodes

    c.   Deploy multiple gateway nodes for load balancing and distribution of client services

    d.   Increase the zookeeper instance count to 5 from 3 (default)

        i.   Ease in maintenance

        ii.   Greater than 5 will slow down the operations due to more voters

4. Components like hive, Ambari, Oozie, and ranger require relational databases as metastore

    a.   Support for Oracle, MySQL, PostgreSQL

    b.   Consider uniformity in databases for easy management

    c.   Provision all metastores on the same server

    d.   Align metastore database management with the usual database administration operational tasks

5. *Capacity planning parameters*

    a. Capacity planning becomes a function of below parameters

        i. Initial data size – historical and current data that will be moved into data lake

        ii. YOY growth – Per year data growth rate

        iii. Compression ratio – the factor by which the data gets compressed

        iv. Replication factor – number of replicas in a cluster

    b. Higher replication factor impacts query performance and data availability. A replication factor of 3 is an optimum number that can balance availability with performance.

    c. Measurement of compression factor varies by data types.

    d. Intermediate data size – Hadoop creates multiple temporary files during intermediate stages. Temp data size accounts for 30-40% of raw data size.

    e. Total storage required

    [(initial data size + YOY growth + intermediate data size) * replication factor * 1.2]

_____

_____

_____

_____

(compression factor)

Note - *1.2 – random buffer factor to account for HDFS storage

    f.   Storm and Kafka

        i.   Storm is compute bound, while Kafka is disk bound

           1.   For storm monitoring, set alerts for capacity, latency, and failed event count

           2.   For Kafka monitoring, set alerts for available disk space and lag between reads and writes

        ii.   If using Kafka, you must plan Kafka log retention period of 2-3 days

6.   *Provisioning and deployment*

    a.   Automate the provisioning and deployment processes through chef, puppet, jenkins, ansible, or cfengine

    b.   Encourage the use of provisioning through management consoles like Ambari or cloudera manager

    c.   After node addition, make sure you re-balance HDFS in an operational window to bring down node threshold levels

7.   *Manage active operations*

    a.   Tune the heap size (~200 bytes per object) as the cluster grows

    b.   Use parallel garbage collection

    c.   Set high availability for metastores, namenode, and security components

    d. Use Ambari to monitor HDFS disk usage, DataNodes, cluster load, CPU usage, and others

    e. If the application uses Hbase, monitor the below metrics

        i. callQueueLength

        ii. memstore size

        iii. compaction queue size

        iv. slowHLogAppendCount

        v. GCTime, CPU Load, CPU Allocation, IOPS

    f. Key logs for troubleshooting

        i. HDFS audit log

        ii. Component logs (/var/log/Hadoop*)

        iii. Application logs (/app-logs/)

        iv. Hive logs (/tmp/<user>/hive.log)

Conclusion

A smooth and stable operations strategy forms the backbone of data lake journey. It gives the confidence to the data and analytics community to come onboard and start playing around with data without any nuisance. A broken scheme to operationalize data lake poses a tough challenge for data governance council and business leadership to realize the essence of data democratization.

Index

© Saurabh Gupta, Venkata Giri 2018
S. Gupta and V. Giri, *Practical Enterprise Data Lake Insights*,
https://doi.org/10.1007/978-1-4842-3522-5

Printed in the United States
By Bookmasters